Alfred Payson Gage

Laboratory Exercises in Physics

Alfred Payson Gage

Laboratory Exercises in Physics

ISBN/EAN: 9783743417441

Manufactured in Europe, USA, Canada, Australia, Japa

Cover: Foto ©berggeist007 / pixelio.de

Manufactured and distributed by brebook publishing software
(www.brebook.com)

Alfred Payson Gage

Laboratory Exercises in Physics

LABORATORY EXERCISES

IN

PHYSICS.

BY

ALFRED P GAGE, A.M.,

AUTHOR OF "ELEMENTS OF PHYSICS" AND "PHYSICAL TECHNICS."

BOSTON:

PUBLISHED BY THE AUTHOR.

1884.

.

J. S. CUSHING & CO., PRINTERS, 115 HIGH STREET, BOSTON.

PREFACE.

IT has been our aim to collate in this volume something of value to every teacher of Physical Science. For valuable suggestions and aid in our efforts, we are deeply indebted to Prof. W. LeConte Stevens of Packer Institute, Brooklyn, N.Y.: Prof. M. B. Crawford, Wesleyan University, Middletown, Conn., and Rev. J. G. Griffin of Ottawa College, Canada. Mr. Arthur W. Goodspeed of Harvard University prepared the key to the solution of problems. Messrs. J. S. Cushing & Co., Boston, are entitled to the credit for the excellence of the typography, and Messrs. Berwick & Smith, Boston, for the presswork.

AUTHOR.

CONTENTS.

———◦◦◦———

PART I.

LABORATORY EXERCISES.

PROPERTIES OF MATTER.—DYNAMICS.

GRAPHICAL METHOD OF REPRESENTING VARIABLE QUANTITIES.

SUPPOSE that we have two quantities, x and y, so related to each other that any change in one alters the other; for example, let x represent the interest, compound or simple, for a term of y years. Take a piece of engineer's paper, Fig. 1, divided into squares by equidistant vertical and horizontal lines. Select one of each of these lines to start from. The vertical line is called the *axis of Y*, and the horizontal line the *axis of X*, and their intersection the *origin*. Take point a as the origin, and let the horizontal spaces to the right represent the number of years, and the vertical spaces multiples of the first year's interest; then points a', a'', a''', etc., represent the interest at the end of the first, second, third, etc., years.

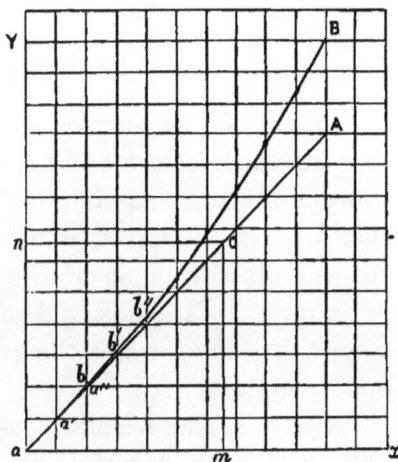

Fig. 1.

Connect these points by a straight line aA. Now, if we take any point, as c, in this line, and connect it with the axis Y by a horizontal line nc, and with the axis X by vertical line mc, the former will represent the time and the latter the interest which has accrued at that time. The line *nc* is called the *abscissa*, and the line *mc* the *ordinate* of the point *c*. In a similar manner are found points *b*, *b'*, *b''*, etc., representing the *compound* interest at the end of the second, third, and fourth years. Having connected these points by the curved line *aB*, the ordinate of any point in this line will represent the compound interest which has accumulated at the time represented by its abscissa. What does the point *A* represent? the point *B*? the line *AB*?

EXPERIMENT 1. Construct a number of curves representing familiar phenomena, as the changes of temperature during the day or year, barometric changes, the velocity of a falling body, the declination of the magnetic needle at New York City (see p. 83) from 1680 to 1880, volumes of a given amount of air corresponding to different pressures, etc.

CRYSTALLOGRAPHY.

EXP. 2. Make saturated solutions of various substances, such as ammonium chloride, ammónium oxalate, potassium nitrate, cuprum nitrate, potassium bichromate, ferrum sulphate, barium chloride, urea dissolved in alcohol, etc., and flow slips of well-cleaned glass with each of these solutions. Allow them to drain for a few seconds, and then, with a microscope or common magnifying glass, watch the growth of the crystals as they form upon the glass.

EXP. 3. If the teacher possesses a porte-lumière, or a stereopticon, he should project the above on a screen by using the wet slips of glass as he would use the ordinary stereopticon slides. If the growth of crystals is slow, as is apt to be the case when the liquids are not fully saturated, it will be well to warm the slips of glass by waving them over a Bunsen or alcohol lamp flame, and then pour the solutions upon the warm glass.

It would be well to encourage pupils to collect cabinets of crystals. The crystals should be preserved in small, well-stoppered bottles or test-tubes. Single crystals may be mounted on heads of pins, the points of the pins being thrust into cork so as to hang from the same inside the bottles.

EXP. 4. Examine with a common magnifying glass a freshly broken surface of cast iron, and observe the crystalline appearance of the fracture.

EXP. 5. Make a cold saturated solution of table salt, and allow it to stand several days. As the water evaporates, small crystals of salt will be formed. *Make drawings of crystals of all the substances used.*

VISCOSITY.

EXP. 6. (By Sir Wm. Thomson.) Take a cake of shoemaker's wax, 18 inches in diameter and 3 inches thick, and place it in a shallow cylindrical glass vessel. Below the cake place a number of corks, and on top of the cake some lead bullets. Fill the glass vessel with water to prevent great variations in temperature. In about a year's time the corks will float up through the wax to the top, while the bullets sink to the bottom, showing the *viscous* nature of the wax.

EXP. 7. Take a strip of sheet lead about 40^{cm} long and 2^{cm} wide, and attach to one end by means of a clamp of some kind a weight of about 300^{g}. Support the whole in a vertical position by means of another clamp applied to the upper extremity, and note at regular intervals of time, by means of a fixed scale placed beside it, the elongation which has taken place. Draw curves of viscosity, the ordinates marking the elongation and the abscissas the units of time.

EXP. 8. Determine the viscosity of glass. Get a glassblower to make a coil about 6^{cm} in diameter and 20^{cm} long from a piece of glass tubing about 1.5^{m} long and 6^{mm} in diameter. Suspend the coil in a vertical position, and attach to the lower extremity a weight of about 20^{g}.

EXP. 9. Determine the viscosity of wires of different metals wound into coils, and draw curves of viscosity for each.

SOLUBILITY.

EXP. 10. Compare the solubility in cold and in hot water of alum, saltpeter, common table salt, white vitriol (zinc sulphate), etc. In each case take 10^{cc} of cold water in a test-tube; pulverize the solid to be tested, weigh out 50^g of it, place a small quantity in the test-tube, cover the mouth with a finger, shake well, and observe the rapidity with which it is dissolved. Continue to add small quantities at a time (smaller as it approaches the saturated point), as long as it is dissolved; and, when saturated, weigh the remaining solid. Its weight $= a$ grams. Then $\dfrac{50 - a}{10} = x$, the solubility of the given substance in cold water.

EXP. 11. Next suspend the test-tube in boiling water, and continue adding small quantities of the substance till saturated. Weigh the remaining solid; its weight $= b$. Then $\dfrac{50 - b}{10} = y$, the solubility of the given substance in water almost boiling. Prepare blanks, and record your results as follows : —

Name of Substance.	Solubility in Water.	
	Cold.	Almost Boiling.
Alum	0.15	3.6

The *solubility* of a substance is the weight in grams of the solid which 1 gram (1^{cc} when water is used) of the solvent requires to form a saturated solution. In stating the solubility of a substance, ought the temperature of the solvent to be given? What weight of alum will 40^{cc} of cold water dissolve?

What weight will the same quantity of hot water dissolve? How does the solubility of alum in hot and cold water compare?

Exp. 12. Ascertain approximately the solubility of the substances used in the last experiment in cold alcohol.

Exp. 13. Put a granule of gum mastic in a test-tube containing water, and shake. Do the same with granules of iodine and analine.

Exp. 14. Repeat the last experiment, using alcohol as a solvent.

Exp. 15. Pour the solution of mastic obtained in the last experiment into a tumbler of water.

Exp. 16. To 4^{cc} of a concentrated solution of sodium sulphate add 2^{cc} of alcohol.

In which are organic substances — *i.e.*, substances of animal or vegetable origin — more commonly soluble, in water or in organic solvents such as alcohol? Can one liquid diminish the solvent power of another?

ABSORPTION AND DIFFUSION.

Exp. 17. Take about 1^{cc} of ammonia water in a test-tube, and immerse the lower end of the tube in hot water; in about a minute close the mouth of the tube with the thumb, invert it in a vessel of cold water, and remove the thumb.

Exp. 18. Make a paste of plaster of Paris about 1^{cm} in depth. Take a glass tube 20^{cm} long and 2^{cm} in diameter, and thrust one end vertically into the paste, and hold it there until the paste hardens. Allow the tube to stand for a day or more to allow the excess of water in the plaster plug to evaporate. Hold the tube vertically, with the plugged end upward, and introduce a rubber tube connected with a gas jet, and fill with illuminating gas. Thrust the open end just beneath the surface of water, and hold it there for a few minutes. The water will gradually rise in the tube in consequence of the osmose of the gases.

Exp. 19. Pour into a saucer about 5^{cc} of ether or bisulphide of carbon, and notice the rapidity with which its vapor diffuses through the air in a room, its presence being recognized by the sense of smell. The molecules issue from the bottle with great velocity; and, if their progress were not interrupted by striking against the air particles, the room would be instantaneously permeated by the odor.

DENSITY, ETC.

Exp. 20. Prepare a concentrated solution of common salt, sulphate of soda, or sulphate of zinc. Introduce first into pure water, then into the solution a wooden demonstration hydrometer like that described in § 64, Physics, and, noting the depths to which it sinks in each liquid, calculate from the data obtained the density of the solution. In doing this, the pupil will intuitively grasp the philosophy of the hydrometer.

Exp. 21. Place an ordinary heavy-liquid hydrometer in water; then add gradually (as fast as it will dissolve) powdered sulphate of soda, stirring with a glass rod, and note from time to time the density of the liquid. Draw a curve of density representing the density by abscissas, and the number of grams of solid dissolved by ordinates.

Exp. 22. Place a light-liquid hydrometer in cold water, then in water at a temperature of about 80° C., and note the density of the two waters. Why must the standard of specific gravity be given at a definite temperature?

Exp. 23. Measure the capacity of some small cavity. First weigh the article containing the cavity, then weigh the same with the cavity filled with mercury, divide the difference between the two weights by the specific gravity of mercury, and the quotient will be the capacity in cubic centimeters.

Exp. 24. According to the principle of parallel forces, the two arms of a balance beam ought to be precisely equal; otherwise, unequal weights will be required to produce equilibrium. Test your laboratory balances by placing weights in the two pans until the beam becomes horizontal. Then interchange the contents of the pans; if the beam remains horizontal, the arms are equal, otherwise it will descend on the side of the longer arm.

Exp. 25. *Paradox.* Take a strip of tin 50^{cm} in length and 6^{cm} in width, and bend it into the form of a circular hoop; solder the two ends together. At some point in the interior of the hoop solder a lump of lead weighing about half a pound. This hoop may now be placed upon an inclined plane in such a position that it will *apparently* roll up hill.

Exp. 26. Bore holes about 2^{mm} in diameter with the point of a pen-knife blade in the opposite ends of a hen's egg; blow the contents out. Drop pulverized rosin through the hole in the large end so as to cover the interior surface of the small end; then pour melted lead through the same hole, so as to " load " the small end. In whatever way you place the shell, it will stand on the small end.

Exp. 27. Take a long glass tube (the longer the better), closed at one end with a tight-fitting cork, fill it with water, and suspend it in a vertical position by a light spiral spring from the ceiling. Suspend at the top of the water column a number of bullets attached to the tube by a thread. With a flame, burn the thread; during the descent of the bullets through the water, the spring contracts and the tube rises. Account for this phenomenon, and make a practical application of the principle involved. When any portion of our atmosphere ascends, in consequence of a denser portion descending, how will the pressure of the lower strata be affected?

Fig. 2.

Exp. 28. Prepare a V-shaped bar like that shown in Fig. 2, the bar *AC* being about 3 feet long; place it so that the end will overlap the table two or three inches, and hang a heavy weight or a pail of water on the hook *B*, and the whole will be supported. Rock the weight back and forth by raising the end *C* and allowing it to fall. What kind of equilibrium is this? Remove the weight, and the bar falls to the floor. Why?

Exp. 29. In the small end of an egg make a hole about 2^{mm} in diameter, and place it, the small end downward, in a wineglass, so that the end of the egg will be within 1^{mm} of the bottom of the glass. Place the whole under the receiver of an airpump, and exhaust the air. The air which is contained in a small sack at the large end of the egg will expand and expel some of the contents of the egg. But, on re-admission of air to the receiver, the pressure of the air will drive the fluid back into the shell.

EXPERIMENTS WITH EIGHT-IN-ONE APPARATUS.

Exp. 30. Insert the stopper a (Fig. 3) in the base, and remove the caps b, c, d, and e, and fill the cup f with water so that the liquid surface will be above the bend g, and *the lateral pressure of liquids* will be shown by the issue of liquids from the side orifices. (It may be necessary to remove temporarily the cap from h to allow the air in the tube to escape.)

Exp. 31. That *pressure increases with the depth* is shown by the increase of velocity of the streams as the depth increases.

Exp. 32. Remove the stopper a, and the liquid ceases to flow from the side orifices, showing that *during the free fall of liquids there is no lateral pressure.*

Exp. 33. Replace the stopper a, and the caps on b, c, d, and e, and remove the cap from h, and connect with this tube, by means of a rubber connector, a glass tube i; and, elevating the latter at various angles, *the exact paths of projectiles* at these angles are shown.

Exp. 34. Remove the stopper a and the cap from j, and close all the other orifices; connect with the tube j a glass tube k, the lower end of which dips into a vessel of liquid m, and this liquid will be drawn up the tube, illustrating the action of the *Sprengel pump.*

Exp. 35. It will be seen that the receiver f is a *Tantalus cup*, and that here it is turned to a practical use, inasmuch as it will not suffer the liquid to flow until the vessel is full, and all is ready for the experiment.

Exp. 36. It is evident that, when the cap is removed from h, and the tube i is suitably elevated, the instrument becomes a *siphon-fountain*.

Exp. 37. The fountain may easily be made to represent an *intermittent spring or fountain* by placing above the receiver a large vessel of water n, from which liquid is siphoned into the receiver f. The siphon delivery being smaller than that of the Tantalus tube, it is evident that the fountain will operate intermittently and at regular periods, inasmuch as the liquid will not flow from the Tantalus cup until it is filled to the level of the bend, and will then flow until the cup is empty.

Exp. 38. Place the apparatus from 8 to 12 ft. above the ground, remove the caps b, c, d, and e, fill the cup f with water, and note the maximum horizontal distance, measured on the ground, which each stream attains. Draw a curve of pressure, the ordinates representing the distance of each orifice below the bend g, and the abscissas the horizontal distances attained respectively by the streams. Draw a curve of velocity on the principle that the velocity varies as the square root of the pressure or *head* of water, the ordinates representing the velocity and the abscissas the pressure at the several orifices.

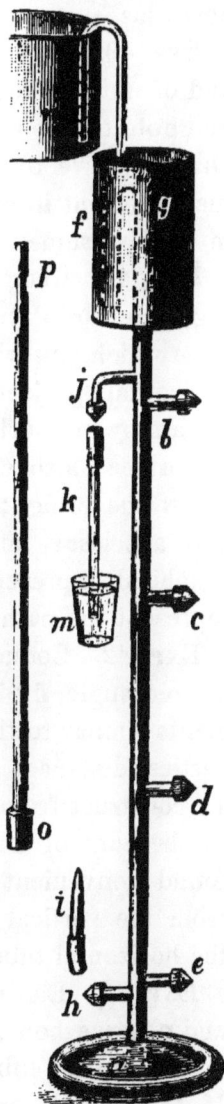

Fig. 3.

Exp. 39. Stop up all the orifices but one, and note the number of seconds it takes to empty the cup through that orifice; do the same with each orifice, and draw a *time curve* representing the time by ordinates and the distance below the bend *g* by abscissas.

Exp. 40. Provide glass tubes 20^{cm}, 40^{cm}, 60^{cm}, and 80^{cm} long, and of uniform bore; connect them successively with the tube *h*, and note the time consumed in emptying the cup through each. Draw a curve of *hydraulic friction*, the abscissas representing the increment in length of tube, and the ordinates the increment in time consumed, or friction.

Exp. 41. Keep the cup constantly full by pouring or siphoning water into it. Close all the orifices save one, the stream from which you wish to represent graphically. Determine the horizontal projection or *random* of suitable points of the stream by measuring perpendicularly from a line let fall from the orifice; also determine the vertical distance of each point below the orifice; then, by means of corresponding ordinates and abscissas, and on a scale of equal parts, construct a graphical representation of the stream. In similar method represent the paths of the streams from the several orifices.

Exp. 42. Connect the glass tube *i* with *h*, elevate it to any desired angle, fix it firmly in this position, and make measurements similar to those in the last experiment, except that the vertical distances are to be measured *upward* from the orifice *h*; construct from the data obtained a graphical representation of the path of a projectile directed at this angle. It will be found convenient to make all the horizontal measurements from the vertical tube *ag*, deducting therefrom the length of the horizontal tube.

Exp. 43. Elevate the tube *i* so that it will be nearly vertical, and observe how much the stream falls short of reaching the orifice of the tube in the cup when the bend *g* is covered. If the stream encountered no resistances from the air, or from friction against the sides of the tube, how high ought it to rise? If to the velocity of efflux from the orifice h there should be added the velocity which is lost in consequence of resistances, how would the sum compare with the velocity which a stone would acquire in falling a distance equal to that from the surface of water in the cup to the tube h?

EXPERIMENTS WITH THE "SEVEN-IN-ONE APPARATUS."

EXP. 44. Suspend the instrument, Fig. 4, from some convenient support, and rarefy the air within it by suction with the mouth; a weight of 20 lbs. ought to be easily raised. This weight, added to a probable friction of about 15 lbs., gives 35 lbs. as the unbalanced pressure exerted by the outside air on the piston.

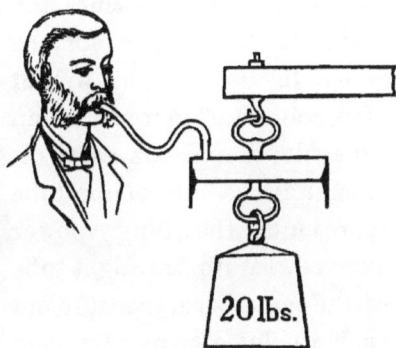

Fig. 4.   Fig. 5.

EXP. 45. Push the piston quite into the cylinder, and close the stop-cock, and let a person grasp each of the handles and attempt to pull the piston out. It is apparent that in his attempt he creates a vacuum, and few will be found strong enough to draw the piston out. This constitutes a most interesting modification of the classical *Magdeburg Hemispheres*, but with this important advantage, that it produces a self-created vacuum (the harder the pull the higher the vacuum), and requires no air-pump.

EXP. 46. Let a person alternately blow and suck air through the rubber tube (Fig. 5), and he will find it difficult to resist the forces (what forces?) tending to move the piston.

EXP. 47. Does the weight which is raised correctly represent the lung-power exerted, or is it a case similar to that in which a force is applied to the long arm of a lever?

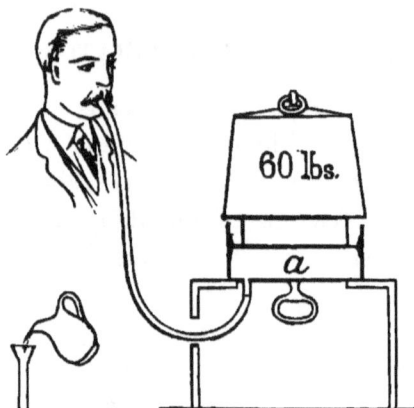

Fig. 6.

Take a glass tube about 6^{dm} long and 5^{mm} bore, bend it into a U-shape; pour into it mercury so that it will stand at a depth of about 15^{cm} in both arms; blow into one arm, and the mercury will rise in the other arm. Measure the hight of the upper surface above the lower; the weight of a column of mercury of an equal depth and the same diameter as the bore of the tube represents the lung power exerted. With a straight tube of the same bore, measure out such a volume of mercury and weigh it.

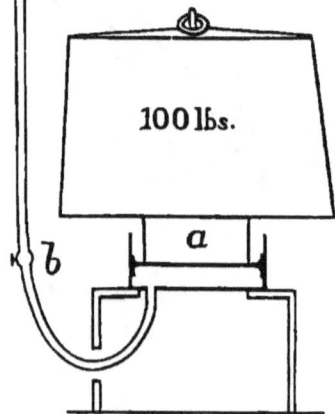

Fig. 7.

EXP. 48. Remove the handle from the piston, invert the instrument, place on the piston a block of wood, as in Fig. 6, and on the block a weight, and support the whole on a box. Blow into the instrument, and a heavy weight may be raised. The instrument thus becomes a *pneumatic bellows*.

EXP. 49. Connect with the rubber tube by means of the stopcock another rubber tube, so that the whole length shall be

12 feet (Fig. 7). Elevate the tube, insert in the upper extremity the funnel-shaped brass mouth-piece, and pour water
down the tube, and a heavy weight may be raised by hydrostatic
pressure, and thus the instrument becomes a *hydrostatic bellows*.

TRIBOMETRIC MEASUREMENTS.

Provide an oak plank 2^m long, 25^{cm} wide, 5^{cm} thick, planed
smoothly on one side; also several sleighs with smooth sur-
faces, as follows: A, an oak plank $14^{cm} \times 14^{cm} \times 2.5^{cm}$; B,
ditto, $28^{cm} \times 14^{cm} \times 2.5^{cm}$; C, a pine plank of same dimensions
as A. Attach a pulley to the middle of one end of the plank in
such a manner that a string fastened to one end of a sleigh and
passing over the groove of the pulley will be parallel with the
surface of the plank. Provide also two pails and a half bushel
of sand.

Exp. 50. Place the plank in a horizontal position, about 2^m
above the floor, and on it sleigh A, near the end opposite the
pulley. Place one of the pails on the sleigh containing sufficient
sand to cause, together with the weight of the plank and pail, a
pressure between the sleigh and plank of (say) 10 lbs. Carry
the string attached to the sleigh over the pulley, and from the
free end suspend the other pail. Place sand in this pail, reg-
ulating the amount, so that, on starting the sleigh with a light
blow, it will move along the plank with approximately uniform
velocity. The weight of this pail, together with the sand, is the
measure of the friction F caused by the given pressure P.

The ratio of the friction to the pressure is called the coefficient of friction; i.e., $\frac{F}{P} = C$, the coefficient of friction.

Exp. 51. Make $P = 20$ lbs., and ascertain F by experiment.
Find the coefficient of friction, and compare it with that obtained
in the preceding experiment.

Exp. 52. Repeat the last experiment, using sleigh B, making
P the same as before. How does the amount of surface in contact affect the amount of friction?

Exp. 53. P remaining the same, find the coefficients of friction when the fibres of the sleigh are parallel to the motion, and when they are perpendicular to the motion.

Exp. 54. With $P = 20$ lbs., and using sleigh C, ascertain the coefficient of friction between oak and pine.

Exp. 55. Ascertain the friction of motion and of repose, *i.e.*, the value of F necessary to keep up uniform motion, and the value of F just sufficient to overcome the state of rest.

Exp. 56. Make the suspended weight great enough so as to produce a slightly accelerated motion, and determine whether the acceleration is uniform. If it is uniform, we must conclude that the amount of friction is independent of velocity.

SIPHON BAROMETER.

Exp. 57. With the blow-pipe flame close one end of a glass tube whose internal diameter is 6^{mm} or 7^{mm}, and length about 1.20^m. Apply the flame so as to make a slight constriction about 20^{cm} from the open end; then about 6^{cm} or 7^{cm} farther, so as to bend the tube into U-form, the open arm being parallel to the closed arm, and distant from it 2^{cm}. If this be filled with mercury, with care lest any air-bubbles remain entangled, the difference of level *ef* between the two columns will be about 76^{cm}, being greater or less in proportion to the atmospheric pressure.

Upon a strip of wood whose length *bc* is 1^m, and breadth *dc* is 5^{cm}, fasten a yard-stick *eg* or a metre-stick; the latter is preferable, and should be placed midway between the two edges of the strip. Any smooth strip of wood will do, if cut squarely off at the bottom *e*, and a mark *f* be drawn across at a distance of 76^{cm} from *e*. On each side of this mark the edge may be graduated in millimeters for a distance of 5^{cm} or 6^{cm}.

Fig. 8.

Place the glass tube against the broad strip so that its lower curved portion may enclose the graduated strip. Secure it loosely with little loops of tinned sheet-iron, h, i, j, k, so that the tube may slide easily past the fixed graduated strip. At the lower end fasten a little block of hard wood bb', projecting enough to form a ledge on which the bend of the glass may rest. Through a hole in this ledge pass a tight screw, with milled head and smoothly-rounded end. The block serves as a nut, and the bend of the glass rests on the rounded end of the screw. The glass may be therefore lifted or depressed by turning the screw.

Adjust the quantity of mercury so that the level in the open arm shall be tangent to the line forming the bottom of the graduated strip at e. This line may be extended across el, on paper, the lower half em being made black with ink, the upper half ek white, so that the hight of the curved surface of mercury at e can be accurately adjusted against a sharply-defined background.

Finally, a screw-eye at g may be inserted, so that the instrument may hang vertically.

If the atmospheric pressure should decrease, the column at f falls and at e rises. By a few leftward turns of the screw, the tube and its contents sink down until e is brought opposite the fixed zero-point. If the pressure increases, a few rightward turns bring e up to the zero-point. After adjusting at zero-point, the hight of the barometric column is read in the neighborhood of f.

HEAT.

Exp. 58. *Conduction.* Lay a thin gelatine card on the palm of the hand. The card being a poor conductor of heat, the lower surface will become more heated than the upper; consequently it will expand more, and bend so that the ends will meet. If it is placed upon a cold surface it will bend in the opposite direction. Glass (especially a variety called lime-glass) ware, when suddenly subjected to great heat, tends to bend in the same manner, but, being only slightly flexible, it is liable to crack, especially if it is thick.

Exp. 59. Wrap the bulb of the thermometer in muslin, and dip it into ether; allow it to remain until the mercury has acquired the temperature of the liquid, then take it out and note the fall of temperature.

Exp. 60. Take a deep beaker of about 800^{cc} capacity three-fourths full of cold water. Pour gently on its surface linseed oil at about 110° C., to the depth of about one inch, and suspend several thermometer bulbs at different depths in the liquid, and notice the slowness with which heat is conducted through the liquid.

Exp. 61. Cut off the nose of a glass funnel of about 800^{cc} capacity. Pass the stem of an air thermometer (having a small bulb) down through the funnel, leaving the top of the bulb about five-sixteenths of an inch below the surface of a card laid temporarily across the edge of the funnel. Fill the space between the neck of the funnel and the stem of the thermometer with sealing-wax, so as to become water-tight. Support the whole in a ring of a ring stand. Partially fill the stem of the thermometer with water colored with ink, and pour cold water into the funnel until its surface is one-fourth of an inch above the bulb. Then carefully pour ether upon the surface of the water until the funnel is full, being careful that none runs down the side of the funnel. Ignite the ether with a match, and notice the slowness with which the heat from a very hot fire penetrates the quarter of an inch depth of water, so as to affect the sensitive thermometer.

Exp. 62. *Convection.* Fill a thin glass flask of about 400^{cc} capacity with hot water deeply colored with ink. Stopper the flask with a cork pierced with a glass tube having a bore of about 5^{mm} diameter. Close the exposed end of the tube with a finger, and thrust the flask to the bottom of a tub or pail (preferably a deep glass jar) filled with cold water, and withdraw the finger. A stream of colored liquid will ascend for a long time from the flask to the surface of the water in the larger vessel.

Exp. 63. Fill a glass flask as before, invert, thrust the neck into water, and withdraw the cork. No convection takes place downward.

Exp. 64. With a porte-lumière and lens, project a large circle of light on a distant screen, and suspend in the path of the beam, beyond the focus of the lens, a red-hot metallic ball or a candle flame.

Exp. 65. Lay a block of ice across the back of two chairs, and over it pass a piece of fine iron wire, the ends of which have been twisted together. From the wire suspend as great a weight as the wire will support, say 25 to 50 lbs. It is evident that since the extent of ice surface on which a fine wire will press is small, the pressure *per square inch* on the ice must be very great. The consequence is that just beneath the wire the ice is melted, and the wire drops down a little. As soon as the wire falls, however, the water about it is relieved from pressure, and immediately freezes. In a short time, therefore, the wire passes completely through the ice.

Exp. 66. Take 200^g of fine dry ice chips or snow at $0°$ C., and an equal weight of water at $80°$ C.; pour the latter upon the former, and with a glass or wooden rod stir and melt the solid as quickly as possible; and as soon as all is melted, take the temperature. The experiment should be conducted in a room whose temperature is as nearly 0° C. as possible.

Exp. 67. Take about a tablespoonful of water in a test-tube, twist around the tube near the bottom a wire for a handle, surround the tube with a freezing mixture, and freeze the water ; or a lump of ice may be dropped into the tube, and a pebble-stone or coil of wire placed on top of it to keep it down. Then pour cold water into the tube, nearly filling it, and hold the tube in a flame (as in Exp. 5, p. 143, Physics), and boil the water at the top without melting the ice.

Exp. 68. Place a beaker of water at about 40° C. under the receiver of an air-pump, and exhaust the air, and cause the water to boil.

Exp. 69. Repeat the last experiment, causing the water to boil for about ten minutes ; at the same time expose to the temperature of the same room in another beaker an equal quantity of water at the same temperature. At the end of the operation take the temperature of both waters. Account for the difference in temperature.

Exp. 70. Do up a piece of writing-paper into a cone shape, gumming the edges smoothly down. Fill it with water, and place it in a loop of wire, and hold it in a Bunsen or spirit flame, and boil the water. The paper will not be charred, for a temperature of 100° C. is not sufficient.

Exp. 71. Paste a strip of paper smoothly around a cylinder of brass or copper, and another strip around a wooden cylinder. Hold them in a Bunsen or spirit flame. The paper on the wooden cylinder will soon become charred, but the paper on the metallic cylinder will not char for several minutes because the metal conducts the heat away so rapidly from the paper that it is not readily raised to the ignition point.

Exp. 72. Take two air thermometers, the bulb of one of them blackened with soot from a candle flame. Let the liquids in the two stems stand at the same hight. Expose each bulb for the same length of time to the sun's rays.

Exp. 73. Repeat the last experiment. except that a convex lens be interposed, and the bulbs placed in the focus of the rays.

Exp. 74. Take two thermometers like those used in the last two experiments, and fill the bulbs and a portion of the stems of each with hot water, and set them, stems upward, in a cool place, and observe by the fall of liquid in the stems which cools more rapidly.

Exp. 75. Partially fill a bladder or the spherical rubber ball furnished for pneumatic experiments with cold air, and place it near a hot stove for a time.

Exp. 76. Fill a test-tube of about 2^{cm} diameter with water. Insert, water-tight, a stopper pierced with a small glass tube, crowd the stopper into the test-tube so that the water will rise in the tube four or five inches. Surround the test-tube with a freezing mixture, and watch for the maximum density of the water, which is reached when the water in the tube reaches its lowest point. Subsequently the water will rise in the tube. This phenomenon is best observed through a telescope a short distance away. Or it may be projected by a porte-lumière on a screen.

ELECTRICITY.

INTRODUCTORY EXPERIMENTS.

EXPERIMENT 77. Place a (tangent)[1] galvanometer so that the needle at both extremities points to zero on the graduated circle ; in other words, so that the coil will lie in the magnetic meridian. Connect the wires leading from a (Daniell or Bunsen) voltaic cell with the screw cups of the galvanometer. This is called. technically, " introducing a galvanometer into the circuit," inasmuch as the galvanometer now forms a part of the circuit, and the current is obliged to pass through it. When the needle *comes to rest*, note the angle of deflection as indicated by the number of degrees to which either extremity of the needle (if there is a difference in the readings, take a mean of the two) points. Observe at which screw cup the current enters the galvanometer, and at which it leaves it, and the direction of the deflection.

EXP. 78. Remove the wires from the screw cups, and, by crossing them (if the wires are not insulated, they should not touch each other), insert each wire in the screw cup opposite the one to which it was before applied. The deflection of the needle is reversed. This shows that the direction of the current through the galvanometer has been reversed. But, by inspection of the circuit, it will be seen that *the current is reversed in only this portion of the circuit.* Can you invent some arrangement by which the current can be reversed in a portion of the circuit more conveniently than by shifting the wires from cup to cup? Such a device would be called a *pole changer* or *commutator.*

---

[1] Two needles are furnished with each galvanometer: one is *astatic*, and when this is used the instrument will be called an *astatic galvanometer ;* when the other needle is used it will be called a *tangent galvanometer.*

Exp. 79. Take a voltaic cell, constructed in the manner described in § 151 of the Physics, first using an unamalgamated zinc. Connect each of the wires attached to the strips of copper and zinc with the galvanometer. Thrust them into the dilute sulphuric acid, holding them quietly in it and at a constant distance apart. Watch for a few minutes the deflection of the needle.

Exp. 80. Repeat the last experiment, using an amalgamated zinc. The deflection diminishes in time, but not so rapidly as in the last experiment.

In the preceding experiment the local currents, established by the impurities in the zinc, did not pass through the main circuit; consequently the main current was very weak. In this experiment, the entire current developed in the battery passes through the main circuit. The *falling-off* of the current which followed the first introduction of the strips into the liquid was due to the *polarization* of the strips.

EFFECTS OF CONDUCTORS OF DIFFERENT LENGTHS, SIZES, Etc.

Exp. 81. Introduce into the circuit with the galvanometer Spool 1 (see Catalogue of Apparatus), containing 32 yards of No. 23 copper wire, in such a manner that the current from the battery will pass through both the galvanometer and the spool; it matters not which it passes through first. To make certain that the connections are all properly made, it is best for a young experimenter mentally to trace the current from the carbon successively through every connection and instrument, back to the zinc of the battery, and see that a suitable path is opened for the current. The screw cups on each side of a given spool are to be used to send a current through that spool.

A deflection smaller than before ensues. The circuit is now 32 yards longer than before, and the result is a weakened current, resulting from increased resistance. Compare the currents in this experiment with the current in Exp. 77, by comparing the tangents of the degrees of deflection as obtained from Table of Tangents, p. 403, Physics. Thus, suppose that the deflection in the first experiment was 84° and in the latter 80°: the tangent of 84°= 9.51 ; of 80°= 5.67. 9.51÷5.67 =1.6 +, *i.e.*, the former current is about 1.6 times the latter.

Exp. 82. Substitute Spool 3, containing 16 yards of No. 23 copper wire, for Spool 1. A larger deflection is obtained than in the former experiment, and you learn that, other things remaining the same, the current varies inversely with the length of the circuit. It is proper to observe here, that if the *entire* circuit were made half as long, in other words, if the entire resistance in the circuit were reduced one-half, the current would be exactly doubled. In this case, only a portion of the circuit is reduced one-half; consequently, the current is not quite doubled.

Exp. 83. Substitute for Spool 3, Spool 2, containing 32 yards of No. 30 copper wire. A smaller deflection is obtained than with Spool 1, which contains the same length and kind of wire, but of greater diameter. Conclusion?

Exp. 84. Obtain the deflection with Spool 4, containing 16 yards of No. 30 copper wire, alone in the circuit, and compare the current with that obtained in Exp. 83.

Exp. 85. Obtain the deflection with Spool 5, containing 16 yards No. 30 German silver wire, alone in the circuit. Compare the current with that obtained in Exp. 84. Conclusion?

INTERNAL RESISTANCE.

Exp. 86. Take a strip of copper and a strip of amalgamated zinc such as used in Exp. 79. Connect the free extremities of the wires with the screw cups of the galvanometer, and introduce the two strips into the liquid, keeping them about half an inch apart. Note the deflection of the needle. Raise the strips half way, and three-fourths way, out of the liquid, noticing the corresponding deflections. As the strips are raised out of the liquid, the size of the liquid conductor between them, in other words, the transverse section of the portion of the liquid between them, is diminished, and the result is an increase of resistance corresponding to the increase of resistance which attends the reduction of size of a solid conductor, as seen above.

Exp. 87. Once more place the strips as at the beginning of the previous experiment, then gradually separate the strips as widely apart as the tumbler will admit, and note the corresponding changes in deflection of the needle. The farther the strips are separated the less the deflection, showing that the effect of increasing the length of a liquid conductor is the same as increasing the length of a solid conductor.

MEASURING RESISTANCES.

Exp. 88. Measure the resistance of the wire on each one of the spools as follows : First, introduce a spool into the circuit with a galvanometer, and get the deflection. Then remove the spool, and introduce in its place the rheostat or set of resistance coils. Place the three switches *A*, *B*, and *C* (Fig. 9) each on the zero butt of its corresponding graduated arc. The circuit will then be closed through this instrument, as may be seen by the deflection of the needle of the galvano-

Fig. 9.

meter. If one of the switches does not at any time rest on a butt, the circuit will be broken. By comparison with the deflection when the rheostat is not in circuit, it will be seen that when all the switches are on the zero butts there is no appreciable resistance introduced into the circuit through the

rheostat. Now you are to aim to obtain the same deflection
of the needle of the galvanometer that you obtained when
the given spool was in circuit. This is done by introducing
through the rheostat a resistance exactly equal to the resistance
of the wire of the spool. You find on the upper surface of the
box which encloses the coils (and thus protects them from in-
jury) three graduated arcs, one extending from 0.1 ohm to 0.9,
the next from 1 ohm to 9 ohms, and the last from 10 ohms to
100 ohms. As in weighing with a balance beam and a set of
weights of three denominations, you put into one of the pans
weights until you succeed in balancing the article to be weighed,
and then add together the weights of the three denominations
to get the total weight, so you introduce resistances into the
circuit by moving one and another switch up their respective
scales until the required deflection is obtained. Then add the
resistances of the three denominations (corresponding to the
tenths, units, and tens of the decimal system of notation), and
the sum is the resistance of the wire and the spool. This
method of measuring resistance is called the *method by substitution*. It is the most expeditious method, and for many practical purposes gives sufficiently accurate results. It is of course
based upon the assumption that the E.M.F. of the battery
remains constant, — a consummation, as will appear farther on.
rarely *fully* realized.

Exp. 89. Introduce into a voltaic circuit through a rheostat a
known resistance r, and obtain a deflection a. Then introduce
in place of the rheostat a wire whose resistance r' is to be measured, and get deflection a'. Since the tangents of angles of
deflection are proportional to the currents, and currents are
inversely proportional to the resistances, we have

$$\tan a' : \tan a :: r : r' ;$$

whence
$$r' = \frac{r \tan a}{\tan a'}.$$

From this formula compute the resistance of the wire. Verify
the result by the " method by substitution."

Exp. 90. Measure the resistance of Spool 4, containing 16 yards of No. 30 copper wire; measure also the resistance of Spool 2, which contains the same kind of wire; calculate the length of the wire on the latter spool as follows:

$$r : r' :: l : l'. \text{ whence } l' = \frac{r'l}{r},$$

in which l is the length of the wire in Spool 4, and r its resistance, r' the resistance of Spool 2, and l' the length of wire to be found.

Exp. 91. Connect in series from five to ten Bunsen cells, and introduce into the circuit with the battery through a rheostat a resistance of (say) 50 ohms. Note the deflection. Remove the rheostat, insert the two poles in metallic handles such as are used in giving shocks. Let a person moisten his hands with a solution of salt in water, and grasp tightly the handles. Note the deflection, and by means of the formula given in Exp. 89, calculate the resistance offered by the person's body.

Exp. 92. *Measure the resistance of a galvanometer* as follows. Introduce another galvanometer into the circuit with the galvanometer whose resistance is to be measured, and note the deflection of the needle of the former galvanometer. Substitute for the latter galvanometer the rheostat, and measure the resistance in the same manner as you would measure any other resistance.

Exp. 93. Measure the resistances of a voltameter, a piece of platinum wire, electro-magnets, either separate from other apparatus or such as constitute parts of apparatus, such as the electro-magnet of a telegraph sounder or a relay.

Exp. 94. Take about 1^m of No. 30 iron wire. wind it into a coil about 1^{cm} in diameter, and as close as possible without allowing the turns to touch one another. Introduce the coil into a circuit with a galvanometer, and note the deflection. Then heat the coil very hot by applying Bunsen flames along its length, and note the diminution of current owing to increased resistance.

MEASUREMENT OF SPECIFIC RESISTANCES.

Exp. 95. Take two wires of the same size and different material ; e.g., copper and platinum, or copper and iron ; get the deflection with one of them (e.g., the iron) in circuit, then replace it by the other, and carry one electrode of the battery along its length until the same deflection is obtained. The ratio between their lengths will express the ratio between their specific resistances.

Exp. 96. Take two wires of different substances and different sizes ; determine the lengths which have equal resistances. Find, by a wire gauge, or micrometer screw gauge, the diameters of the wires, and then ascertain the ratio of their specific resistances by the formula

$$s \, \frac{l}{q} = s' \frac{l'}{q'},$$

in which s and s' represent their respective specific resistances, l and l' their lengths, and q and q' the areas of their cross sections or squares of diameters.

EFFECT OF CORRODED CONNECTIONS.

Exp. 97. Find (they are usually not difficult to find) some piece of electrical apparatus which has a rusty or corroded connection, such as rusty screws of screw-cups, either connected with the battery or other apparatus. If these cannot be found, select a wire which has become corroded at one or both ends by exposure to acid fumes. Introduce this corroded connection into the circuit with the galvanometer, and note the deflection. Now clean all the rusty connections with a fine file, fine sand- or emery-paper, or by scraping them with a knife, and once more send the current through the same apparatus. The deflection is now greater than before. Introduce through a rheostat a resistance sufficient to give the same deflection as at first, and thus measure the resistance caused by the corroded surfaces of contact. *Beware of corroded connections.*

JOINING CELLS IN OPPOSITION.

Exp. 98. Take two cells as nearly alike as possible, connect the zincs of each cell with each other by a short copper wire, and the two carbons with the galvanometer. There is either no deflection, or, at most, a very small deflection. This is called *joining cells in opposition*. Here the tendency of one cell to produce a current in a certain direction is counterbalanced by the tendency of the other cell to produce a current in the opposite direction, and no current flows, provided the E.M.F. of both are equal. If the E.M.F. of both are not equal, then there will be a current proportionate to their difference.

ELECTRO-MOTIVE FORCE.

Exp. 99. Connect, in opposition, two cells of the same kind but of different sizes ; *e.g.*, a quart cell with a gallon cell ; or, which amounts to the same thing, lift up the zinc or the carbon, or both, of one of the cells, thereby diminishing the immersed part of one of the cells ; no deflection ensues. It thus appears that a large cell has no greater power to produce a current than a small cell ; *i.e.*, *the E.M.F. of battery cells does not depend upon their size*.

Exp. 100. Connect, in opposition, two cells of different kinds ; *e.g.*, a Bunsen cell and a Daniell cell ; a deflection ensues, showing that the E.M.F. of the two cells is different, and that *the E.M.F. of voltaic cells depends upon the material used*.

MEASUREMENT OF ELECTRO-MOTIVE FORCE.

Exp. 101. Measure the E.M.F. of a voltaic cell as follows : Take, for example, a Bunsen cell, and connect in opposition with it a Daniell cell, and introduce into the circuit with this combination a galvanometer. There will be a deflection of the needle. Add other Daniell cells, one at a time, in series, in opposition to the Bunsen cell, until there is no deflection, or only a slight deflection.   Inasmuch as the E.M.F. of the Daniell cell is (about) 1 volt, the E.M.F. of the Bunsen cell will be approximately as many volts as the number of Daniell cells required to neutralize it.

EXP. 102. *Measure E.M.F. by method of equal deflections,* the standard cell being E, the one to be compared being E'.

Take the deflection of E and call it R; then take the deflection of E' and call it R', adding resistance to make the deflection the same.   Then

$$E' = E\frac{R'}{R}.$$

EXP. 103. *Measure E.M.F. by comparison.*   Call the E.M.F. of two batteries E and E'; join them up successively in circuit with the same galvanometer, and, by varying the resistance, cause them both to give the same deflection; their forces will then be in direct proportion to the *total* resistances in circuit in each case;

or, $$E' = E \times \frac{R'}{R},$$

when R (including that of battery, galvanometer, and the adjustable resistance) represents the resistance with E, and R' with E'.

EXP. 104.   Place the cell whose E.M.F. is to be measured in circuit with a galvanometer; it produces a deflection of d degrees; then add enough resistance r to reduce the deflection to d' degrees, say 10 degrees less than before.   Now substitute the standard (Daniell) cell in the circuit, and adjust the resistances through a rheostat till the deflection is d, as before; then add enough resistance r' to reduce the resistance to d'.   Now, calling E the E.M.F. of the battery to be measured, and E' the E.M.F. of the standard battery,

$$r' : r :: E' : E, \text{ whence } E = \frac{rE'}{r'},$$

since the resistances which will reduce the current equally will be proportional to the electro-motive forces.

MEASUREMENT OF INTERNAL RESISTANCE.

EXP. 105. Measure the internal resistance of battery cells as follows. Take two cells which, connected in opposition as in Exp. 98, will give no current; then introduce a third cell into the circuit with these two cells and a galvanometer, and a deflection will be produced by the latter cell. Note the deflection, remove the pair of cells connected in opposition from the circuit, and in its place introduce the rheostat, and measure the resistance of the pair of cells which has been removed. One-half of this resistance is the resistance of a single cell.

EXP. 106. Put the battery in circuit with galvanometer, and note the deflection. Halve the tangent of the deflection by introducing resistance. The resistance introduced is equal to the original resistance — that of the battery and the galvanometer coil. Deduct from the resistance introduced the resistance of the galvanometer, and the remainder is the resistance of the battery.

EFFECTS OF POLARIZATION.

EXP. 107. Introduce a Bunsen cell, whose zinc is well amalgamated, into circuit with the galvanometer, and watch from the commencement the deflection of the needle for about ten minutes. The deflection will diminish somewhat during the first five minutes, and afterward remain quite constant. A deterioration of current, greater or less, due to polarization during the first few minutes of its use, takes place with the best batteries, and proper corrections should be made for the same in all experiments in electrical measurements.

EXP. 108. Connect in opposition a fresh cell with a cell which has been working from five to ten minutes, introducing a galvanometer into the circuit. A deflection of the needle shows that the E.M.F. of the fresh cell is greater than that of the other.

CURRENT THE SAME AT ALL POINTS OF A CIRCUIT.

Exp. 109. Introduce into circuit a resistance of 52 ohms (about the resistance of four miles of telegraph wire) through a rheostat ; and. between the positive terminal and the rheostat, a galvanometer. Measure the current at this point, the beginning of its journey. Then insert the galvanometer between the rheostat and the negative terminal, and measure the current near the end of its journey. Finally introduce two rheostats into the circuit, and the galvanometer between them, and throw a resistance of 26 ohms from each into the circuit. Compare the results of the three trials. and determine whether they verify Law VI., page 68.

EFFECTS OF DIFFERENT METHODS OF JOINING CELLS.

Exp. 110. Take a single cell, and introduce into its circuit a resistance coil and galvanometer. By means of the coils throw a resistance of (say) 15 ohms into the circuit. Note the deflection. Introduce another similar cell into the circuit connected abreast (see § 183 of the Physics) with the first, still retaining in the circuit the 15 ohms resistance. The deflection is *very slightly* increased if at all, by the introduction of the additional cell. Find and compare the currents in the two cases, assuming that the E.M.F. of the (Bunsen) cell is 1.8 volts, and $r = 0.5$ ohm, and $R = 15$ ohms, disregarding the resistance of the galvanometer. It will be seen, both from the experiment and the calculation, that in this case almost no advantage is gained by using two cells connected abreast instead of one. By joining the two cells abreast, we virtually make one cell of double size, and thereby reduce the internal resistance one-half. But as the internal resistance of a single cell (0.5 ohm) is a small portion of the whole resistance in the circuit, the advantage gained is proportionately small.

Exp. 111. Connect the two cells *tandem*, retaining the external resistance of 15 ohms in the circuit. The deflection is now much greater than when a single cell is used in the circuit. Compare the currents in the two cases ; also, compute by Ohm's law the currents in the two cases. Here the resistance of the battery is double that of a single cell, or the whole resistance of the circuit is increased 0.5 ohm by the introduction of an additional cell. But this is a small portion of the entire resistance, and is, therefore, in this case, of little consequence. On the other hand, the E.M.F. of the battery is doubled. Consequently, the current, as shown both by experiment and calculation, is nearly doubled.

EXP. 112. Introduce a galvanometer into a circuit with a single cell by means of short, thick copper wires, and note the deflection. Then connect two similar cells abreast, and note the deflection. The deflection is much increased by the introduction of the additional cell. Compute the current in each case, assuming that the internal resistance of each cell is 0.5 ohm, and the external resistance is too small to be regarded. In both cases the whole resistance of the circuit is the internal resistance of the battery. By connecting two cells abreast, we reduce this resistance one-half, and consequently double the current, as shown both by experiment and calculation.

EXP. 113. Connect the two cells used in the last experiment *tandem*, and note the deflection. The deflection is no greater (or very little greater) than that obtained with a single cell in circuit. Calculate the current in this case. By connecting the two cells tandem we double the E.M.F. of the battery. This of itself would double the current, but on the other hand we double the *resistance* of the circuit. This would counterbalance the advantage gained by an increased E.M.F.

EXP. 114. Introduce through a rheostat 4 ohms' resistance into a circuit with two Bunsen cells connected abreast. Note the deflection. Then connect the same cells abreast, and note the deflection. Compute the currents in both cases, and determine whether the results of the experiments verify your calculations.

Exp. 115. Introduce through a rheostat a resistance of 2 ohms into a circuit with six Bunsen cells connected *tandem*. Note the deflection. Remove from the circuit five of the cells, and obtain the deflection with a single cell in circuit. Connect the six cells *abreast*, and obtain the deflection. Finally. connect the six cells as a pair of triplets as follows. Take three cells and connect the three zincs with one another; also the three carbons with one another. Connect the other three cells in the same manner. Then connect the zinc of one of the triplets with the carbon of the other triplet. In connecting the triplets with each other, it is immaterial at what point of the zinc combination or the carbon combination the connection is made. For example, the copper wire which is to connect the triplets may be attached to any one of the three zincs, or to the wire which connects the three zincs, for the three zincs in the triplet are to be regarded and treated in every respect as though it were one zinc. The same is true of the carbon triplet. Compute the current in each of the four cases, assuming that the E.M.F. of each cell $= 1.7$ volts, and r of each cell $= 0.5$ ohm, the external resistance in each case being (disregarding the resistance of the galvanometer) 2 ohms. Observe whether the results obtained by calculation are verified approximately by the results obtained by experiment.

Exp. 116. Introduce a voltameter or apparatus for decomposing water into a circuit, and measure its resistance, and determine in what way a battery of two cells (Bunsen) should be connected for electrolytic purposes.

ELECTRO-MAGNETS OF VARYING RESISTANCE.

Exp. 117. Introduce into a circuit. with a single (Bunsen) cell, an electro-magnet wound with very fine wire. Measure the resistance of the electro-magnet; then introduce by means of the rheostat an additional resistance equal to that of the magnet. Note the deflection, and test the strength of the electro-magnet by applying an armature to its poles, and observing the force necessary to pull it off. Then substitute for this electro-magnet an electro-magnet wound with coarse wire, allowing the same resistance through the rheostat to remain in circuit. Test the strength of the electro-magnet as before.

EXP. 118. Introduce the electro-magnet of fine wire into circuit, without any other resistance except that of the battery. Test the strength of the magnet as in the last experiment. Then for this substitute the magnet of coarse wire, and test the strength of this magnet. Do the results of the last two experiments verify the first law of electro-magnets, page 72?

EXP. 119. Place in circuit the same magnet of fine wire as used in the last experiment, with an equal resistance through the rheostat. Introduce an additional cell connected in series with the first. Test the strength of the electro-magnet. Note the deflection produced in the galvanometer introduced into the circuit. Compare the present current with the current when a single cell was in circuit with the same resistances.

EXP. 120. Place in circuit with a Bunsen cell a telegraph key, and a sounder of low (about 3 to 9 ohms) resistance (such as furnished by the Author). The cell will work the sounder strongly. Then introduce into the same circuit with the last a relay of about 25 ohms resistance (such as furnished by the Author). The single cell will probably not work either the relay or sounder, in consequence of the increased resistance. If either works it will be the relay.

EXP. 121. Now introduce two cells connected tandem into circuit with the relay and sounder. The relay will now work, but the sounder probably will not ; or, if the sounder does work, it will be only feebly. In case it does work, introduce sufficient additional resistance into the circuit through a rheostat, so to enfeeble the current that the sounder will not work, but not sufficient to prevent the relay from working. Why will the relay work and not the sounder?

EXP. 122. Take a third voltaic cell, and construct a local circuit, with the sounder in this circuit, inserting the wires of this circuit in the two screw cups of the relay that have not so far been used. Now manipulate the key in the line circuit, and this will work the relay in the same circuit; the relay working will open and close the local circuit, causing the sounder in this circuit to work.

EXP. 123. Introduce, through a rheostat, into the circuit with the relay a resistance (say) of 65 ohms (which is about the resistance of five miles of telegraph wire). Then introduce a sufficient number of cells, connected in series, to work the relay. Compute the current which works the relay, first measuring the resistance of the relay, and assuming that the r of each cell $= 0.5$ ohm, and that the E.M.F. of each cell $= 1.8$ volts.

DIVIDED CIRCUITS.

EXP. 124. Introduce into a circuit with a single cell B (Fig. 10) a galvanometer G, and rheostat R, and through the last a

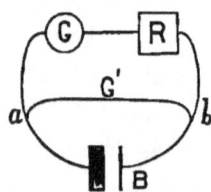
Fig. 10.

resistance of 4 ohms. Note the deflection. Now take a short copper wire and attach it to the two battery wires at any two points, as a and b, by winding several turns around the wire so as to make good connections. If double connectors are used, these connections may be made easily and quickly. Note the deflection, and observe that it is very much less than before. The wire which serves as a bridge between the two battery wires is called a *shunt*. Next cut the shunt wire at some point, as at G', and introduce the galvanometer at that point. Close the circuit at the point from which the galvanometer was removed, by twisting the wires together, or, better, by inserting a double connector. A deflection of the galvanometer shows that a current is passing through the shunt wire. Observe that the deflection is greater than it was when the galvanometer was in circuit with the rheostat. We thus learn that the current becomes divided unequally at one of the two points a or b (according to the direction in which the current is flowing), and the two portions flow through the two paths open to it from that point, *the larger portion flowing through the smaller resistance.* Compare the currents in the two branches and see if the results of the experiment verify the law of divided circuits, page 69.

EXP. 125. Remove the galvanometer from the shunt wire, closing the circuit at that point, and introduce it into the circuit between the battery and point a or b. Observe that the deflection is now larger than it was before the circuit was shunted, as we might expect when we reflect that by introducing the shunt wire we virtually increase the size of the wire in the circuit beyond the points a and b, and thereby reduce the resistance of the circuit.

If a main wire is split or divided into many branches, all of equal resistance, the current will with absolute certainty and perfect exactness divide itself equally among all the branches. This fact renders the distribution of an electric current among any number of electric lamps, so that all shall be equally lighted, perfectly simple and easy, even more easy than the equal distribution of gas among a number of burners. If, however, it is desirable to make an unequal distribution, it may be very easily done by suitably varying the resistance in the different branches.

THREE METHODS OF TRANSFER OF ELECTRIFICATION.

EXP. 126. Charge a Leyden jar in the usual manner; place one ball of a discharger on the outside coating, and bring the other ball on a level with and about 8^{cm} from the ball of the jar; suspend between the last two balls by a silk thread a pith ball covered with tin or gold foil. The pith ball acted on by the electric force will vibrate between the two balls, carrying electricity from one to the other, and thus gradually discharge the jar. A transfer or current of electricity is thus kept up between the two balls in virtue of the motion (caused by the E.M.F.) of the electrified body which *conveys* the electricity as it moves from place to place. This phenomenon may be called a *current of convection*.

" Electricity *carried* through space on a charged body has exactly the same magnetic effect on a stationary magnetic needle as if it had been conducted." — ROWLAND.

EXP. 127. Charge the jar again and establish an electrical connection between the ball of the jar and the outside coating through a discharger. There will be a transference of electricity through the wire of the discharger, but the wire itself does not move. What takes place in the wire is called a *current of conduction*.

When a compound body is decomposed by an electric current, the mode in which the current is transmitted through the electrolyte is called *electrolytic conduction*, and is always associated with a flow of the components of the electrolyte in opposite directions.

EQUIPOTENTIAL LINES AND LINES OF FLOW.

EXP. 128. Take a sheet of tin foil AB about 25^{cm} square, spread out over a sheet of paper; at points x and t apply the

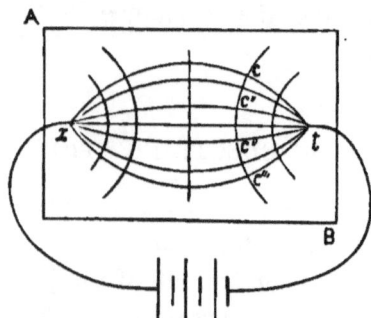

Fig. 11.

two terminals of a battery. Then with two platinum electrodes connected with an astatic galvanometer, feel around for points c, c', c'', etc. (Fig. 11), such that when the platinum terminals are applied at any two of them the galvanometer shows no deflection. Prick through these points into the paper.

Points whose potentials are equal are called *equipotential points*. A line connecting a series of such points, as cc''', is called an *equipotential line*. As the experiment shows and the term implies, there is no current along an equipotential line.

If the direction of the E.M.F. at different points in the field are found (as may be done by suspending a very delicate elongated conductor successively over these points, the axis of the conductor placing itself in the direction of the force at each point), and if a line be drawn so that its direction at every point coincides with the direction of the electric force at that point, such a line is called a *line of force*. It is found that the direction of the E.M.F. is everywhere perpendicular to the equipotential lines, and hence the line of force everywhere cuts these lines at right angles and terminates in the electrified points or centers of force as x and t. By drawing a number of such lines as in Fig. 11, the direction of the force at different parts of the field may be indicated. Since these lines also indicate the direction in which the transfer of electrification takes place, they are called *lines of flow*.

"If from a sheet of indefinite extent we cut off a portion bounded by lines of flow, we shall not affect the electrical distribution, *i.e.*, the forms of the equipotential lines will be the same as before the sheet was cut." See Gordon's "Electricity and Magnetism," pp. 31–41, Vol. II.

INDUCTION COILS.

Exp. 129. Connect the secondary coil of an induction coil with a (tangent) galvanometer (see Fig. 170, Physics), and the primary coil with a battery. Introduce the latter coil into the former, and as soon as the needle becomes quiet, withdraw it, noting the deflections. If the deflections are too large, so as to throw the needle beyond 90°, the deflections may be reduced any desired amount by means of a shunt of suitable resistance. Then, allowing the primary coil to remain within the secondary coil, plunge an iron core within the former, and soon withdraw it, noting both deflections. Finally, introduce both primary coil and its core into the secondary coil, and presently withdraw it, noting the deflections. It will be seen that the core is a very important adjunct to an induction coil.

Exp. 130. Remove the core from the induction coil; let the primary coil remain within the secondary coil, and introduce the former together with the vibratory circuit-breaker into circuit with a single voltaic cell. Connect a pair of handles with the terminals of the secondary coil, and let a person grasp these handles, holding one in each hand. Let another person gradually introduce a core into the primary coil. The intensity of the shock will rapidly increase until it will become unendurable. The intensity of the shock can be governed by the distance the core is inserted in the coil. If the coil without the core should give painful shocks, the primary circuit may be shunted through a rheostat and any desirable resistance.

Exp. 131. Connect, by means of flexible wires several meters long, the secondary coil of a small induction coil with an astatic galvanometer, and turn it end for end in an E. and W. plane, and feeble currents will be induced therein through the influence of the earth's magnetism.

Exp. 132. Connect a telephone receiver with a sensitive long-coil galvanometer, and produce induced currents by pressing the disk of the telephone inward with the finger and allowing it to spring back.

Exp. 133. Shake the hand rapidly with the fingers spread out in front of a Geissler tube illuminated by an induction coil, and the light of the spark, being intermittent, will produce a large number of images.

Exp. 134. Select a Geissler tube contracted along the centre so that the light is reduced to a narrow, bright line. Place the tube in a vertical position, and a bi-sulphide of carbon prism a little distance from it, and project a spectrum of the light upon the screen. The spectrum thus obtained consists of several bright lines characteristic of the gases which they contain.

Exp. 135. Insulate a Leyden jar, and connect the secondary

terminals of a Ruhmkorff coil with the two coatings of the jar. Lead wires of different metals from these coatings so that their extremities may be from one to two inches apart, and observe the character of the spectra produced during the discharges. The spectra produced should be characteristic of the metals used.

Exp. 136. "To show the production of induced currents in a telephone and their physiological effect. Attach the ends of the wires from a Bell telephone to the leg muscles of a frog, and speak in the telephone. The pronunciation of the word 'sucker' causes the leg to move or 'jump,' while 'lie still' has scarcely a sensible effect."

Exp. 137. Take a coil of wire of considerable size, e.g., the primary coil of an induction coil (Fig. 170, Physics), and a soft iron core a little longer than the coil and about one-half inch in diameter, attach to it the poles of a battery of two cells of Bunsen, so connected as to give a strong current. Make and break the circuit by touching and separating the extremities of the wires or by a telegraph key inserted in the circuit, — the key and coil may well be in separate rooms, — and a person placing his ear near the core will hear distinct clicks ("magnetic ticks") every time the circuit is broken and closed. These *sounds proceed from molecular disturbances attending magnetization and demagnetization.* Experiments have determined also, that *iron rods are elongated on being magnetized, and retract on losing magnetism, and that heat is developed in the iron by both processes.*

Exp. 138. Lead the secondary terminals of a Ruhmkorff coil to the opposite coatings of a Leyden jar (the larger the better), and place one ball of a discharger on the outer coating and hold the other ball near the ball of the jar. The discharges will be less frequent as the jar has to be charged between each spark, but the sparks will be much more brilliant and the reports almost deafening. The jar should be placed on an insulating stand.

Instead of a single jar, several connected " in cascade " (see p. 44) may be used. All the jars in the cascade must be insulated.

EXTRA CURRENTS.

Exp. 139. Connect one of the poles of a battery with one extremity of a file as in Fig. 132 of the Physics, and draw the other pole over its roughened surface. Yellow sparks fly from the file. These are incandescent particles of iron projected into the air, caused by the fusion of the metals at the points of contact. Introduce into the circuit a helix and repeat the experiment. Now, in addition to the yellow sparks which fly from the file, there will be seen much brighter and whiter sparks following the electrode as it passes over the surface of the file. These sparks are confined to the surface of the file, and are caused by the passage of the extra currents generated at each break of the circuit (the extra current on " making" gives no spark). These sparks are of the same nature as those which pass between the secondary poles of an induction coil. Finally, insert an iron core in the helix and again repeat the experiment. The sparks produced by the extra currents are still brighter.

ELECTROLYSIS.

Exp. 140. Lay upon a metallic surface — e.g., a strip of brass, copper, or tin — a piece of white paper moistened with a solution of salt, to which a few drops of ferrocyanide of potassium have been added ; connect the negative electrode of a battery with the metal strip ; make a positive electrode of iron by attaching to the copper wire a short piece of iron wire ; move the free end of the iron wire over the paper, as if to write on it, and a blue mark will be made on the paper. See § 243, Physics.

Exp. 141. Dissolve about 1^s of iodide of potassium in 20^{cc} of water ; make a starch paste by boiling a small quantity in a teaspoonful of water ; take as much of the paste as will lie upon a quarter of an inch of the point of a penknife blade, and stir it in the solution. Moisten a piece of white paper with this solution ; lay it upon a metallic surface ; connect one of the copper poles of the battery with the metallic surface, and, with the other pole, write upon the paper as with a pencil.

HEATING EFFECT.

Exp. 142. Provide a glass flask of one liter capacity. Pass water-tight through a cork two No. 16 Kerite covered copper wires, and, close to the cork, attach to one an amalgamated zinc rod, and to the other a carbon rod, the rods being of sufficient length to reach to about half the depth of the flask. Connect the two extremities of the wire, outside the flask, by a platinum wire about 2^{cm} long and size No. 32. Introduce into the flask bichromate of potash solution enough to nearly reach the lower extremities of the rods when the flask is upright. When you would show the heating power of electricity, light gas by means of the heated wire, etc., you have only to invert the flask.

THERMO-ELECTRIC CURRENTS.

Exp. 143. Take a strip of sheet copper about 15 inches long and three-fourths of an inch wide, and a strip of zinc of the same dimensions. Lay them one upon the other, and fold over each end upon itself for about half an inch, and hammer the joints flat, so that they shall hold together quite firmly. Then separate the two strips into a somewhat elliptical or rectangular shape, as shown in Fig. 12. Cut a hole through the center of one of the strips, and pass the wire support of a

Fig. 12.

magnetic needle through it. Place the band in the magnetic meridian parallel with the needle. Direct a flame against one of the junctions, and note the deflection, and determine the direction in which the current traverses the band, *i.e.*, whether the current passes from the heated junction through the copper or the zinc strip.

EXP. 144. Attach to a (astatic) galvanometer a copper and an iron wire. It is well to let the wires rest upon pieces of ice near the galvanometer. Join the free extremities of the wires, and apply a Bunsen or spirit flame at the junction. A current is established, and the E.M.F. continues to increase till the temperature of the junction is about 284° to 300° C. ; then, on raising the temperature higher. it begins to decrease, and, finally, is reversed, as is shown by a reversal of the deflection.

MAGNETISM.

EXP. 145. Heat red-hot an iron ball, and suspend it by a brass or copper wire. Bring a pole of a powerful electro-magnet near to the ball, and the latter will not be attracted. Keeping the magnet in the same position, wait until the ball has cooled, and, when sufficiently reduced in temperature, it will be attracted. This shows that iron cannot be magnetized when red-hot.

EXP. 146. Hold a soft-iron rod about 1^m long in the same position as that taken by a dipping-needle. Suspend a small magnetic needle, and bring it near the end of the poker which points downwards. The N. pole of the needle will be repelled, and its S. pole attracted, showing that this end of the poker is a N. pole. Pass the needle upwards close to the poker, and its S. pole will continue to be attracted until the middle is reached, after passing which the needle will turn round, and its N. pole will be attracted ; and so onwards to the upper end of the poker, which we say has been magnetized by the *inductive action* of the earth.

EXP. 147. *Effect of percussion.* While in the position mentioned in the last experiment, strike the end of the poker several times with a hammer. This percussion will be found to have conferred *coercive force* upon the poker, which now retains its magnetism in any position.

Exp. 148. *The earth's magnetic force directive and not translative.* Float a small magnetic needle, fixed horizontally to a cork, on water. The needle will point nearly N. and S., but will not float towards the north. This is owing to the great distance of the earth's magnetic poles, which are both attracting and repelling the needle. The attraction and repulsion being practically equal, there results only a rotation, and no motion of translation. The two forces act as a *mechanical couple.*

EXPERIMENTS WITH LEYDEN JARS.

Exp. 149. Charge the jar by grasping the outer coating with the hand, and holding the ball about half an inch from the prime conductor. When the sparks cease to pass from the conductor to the ball, the jar is charged.

Exp. 150. Place a charged jar on an insulating support, and touch with a finger the ball, keeping the other hand away from the outer coating. A faint spark will pass from the ball to the finger. Now touch the outer coating, and a faint spark will pass from this coating to the finger. Continue touching alternately the ball and outer coating until the jar is discharged, or as long as a spark can be obtained from either ball or coating.

Exp. 151. Place a Leyden jar on an insulated support, and bring the ball near the prime conductor. Only a few sparks will pass from the conductor to the ball. After turning the machine for a time, discharge the jar, when it will be found that it contained only a slight charge, thus showing that *an insulated jar cannot be heavily charged in the usual manner.*

Exp. 152. Take a strip of leather one-half inch wide and long enough to encircle the jar. Through this strip, at intervals of one inch apart, thrust sharp pointed tacks. Bind the strip around the jar and over its outer coating, so that heads of the tacks will press against the jar. Now place the jar upon an insulating support, and it will be found that it can be easily and heavily charged.

Exp. 153. Take three or four Leyden jars, and place each horizontally on its side, on separate insulating supports. Bring the ball on one of the jars near the conductor of the machine, and the ball of a second jar near the outer coating of the first jar, and so on for the third and fourth jars. Let the outer coating of the last jar be held in the hand or connected with the earth, or. still better, connected by a chain with the negative conductor of the machine. Now, as the machine is worked, each jar will become charged. The interior coating of each jar will be charged with a different kind of electricity from that of the jar which precedes it, and will be less heavily charged than its predecessor. Each jar may be separately discharged in the usual manner, or the whole may be simultaneously discharged by first connecting all interior coatings with one another, and all the exterior coatings with one another; then, with a discharger, discharge the whole as if it were a single jar.

Exp. 154. Charge a jar from the prime conductor of an electrical machine, and, rubbing the ball over the resinous surface of an electrophorus, or a plate of glass covered with shellac, draw a figure upon its surface. Charge another jar from the negative conductor of the machine, and draw another figure across the last figure. Now sift the dust of a mixed powder composed of red lead and sulphur over this surface. The two powders will separate and arrange themselves in beautiful radiations. the red lead along the lines formed by the negative jar, and the sulphur along the lines formed by the positive jar.

Exp. 155. Charge separately, and in the same manner. several Leyden jars. Place them in a row on insulating stands. and connect them in series, after the manner of a voltaic battery whose cells are connected tandem, i.e.. connect the inner coating of the one with the outer coating of the next throughout the series. Then discharge the whole. by connecting the outside coating of the jar at one end of the row with the inside coating of the jar at the other end of the row. Very long and intense sparks are thus obtained.

Exp. 156. Take 20 threads of linen each 12 inches long, and, placing them together and parallel with one another, so as to form a bundle, tie strings around the bundle 1 inch from either end. Suspend the bundle, by means of a short wire, from the prime conductor of the machine. As the machine is worked the threads will recede from each other and form a balloon-shaped figure.

Exp. 157. Let some ingenious boy cut from a cork or pitch a figure of the body of a spider, attaching to it linen threads about 1½ inches long for legs. Suspend the spider from some support by means of a silk thread. Charge two Leyden jars, one with negative and the other with positive electricity. Place the two jars near each other on a table and bring the spider between the two balls. It will be first attracted to one of the balls, then repelled by it and attracted by the other ball, and will continue to vibrate for a considerable time between the two balls, the legs at each end of the swing grasping the ball as if to support itself. At each swing it carries a small charge of electricity from jar to jar, and thereby gradually discharges the two jars.

Exp. 158. Take a warm, dry, wide-mouthed, shallow bell-glass or a glass fruit-dish, and rub the ball (better a point) connected with the prime conductor, while the machine is in operation, over the interior surface of the glass vessel until the interior surface becomes highly excited; then place it so as to cover a handful of pith balls lying upon a table. An animated scene will follow.

Exp. 159. Insulate an electrical machine, and connect the positive and negative conductors by a wire. However rapidly the machine is worked no charge will accumulate, thus showing that *equal quantities of + and − electricity are produced.*

Exp. 160. Steep Swedish filter paper in a mixture of equal volumes of nitric and sulphuric acids, wash in abundance of

water, and dry. Lay it on a sheet of waxed paper and rub
briskly with flannel or silk, and use as an electrophorus.

METAL SCREENS.

Exp. 161. Electrify a glass or sealing-wax rod and interpose
between it and a pith ball a metal screen (*e.g.*, a strip of wire
gauze) connected with the earth ; no induction takes place. In
general, *a body may be protected from inductive influence by
covering it with a metal or wire gauze cage connected to earth.*

"The problem of lightning protection is to produce a space
into which electricity cannot enter. If a hollow shell of copper
were made, for example, a person inside would be perfectly
safe, though all the lightning of the heavens were playing about
him, because the electricity would pass around the outside,
which would be of superior conductivity, instead of leaping
across the space within. So, if a house were enclosed in a
cage of copper, the lightning would pass around the cage instead
of through it. The best method of protecting a house, therefore, would be to erect a central rod on the roof, from which
conductors would pass to the four corners and then down to the
ground. But the rods must not stop there ; they must continue
beneath the house, surrounding it below as above, completely
enclosing the bottom as well as the roof. If it were desired to
make this cage more complete, conductors might be carried
from the central rod down the four sides of the house, as well
as down the corners."— ROWLAND.

SOUND.

VIBRATIONS OF BARS.

EXP. 162. Let a carpenter prepare a pine bar as nearly uniform throughout as possible and 5 feet long, 1 inch wide, and $\frac{3}{16}$ inch thick. Cut it, as nearly as possible, into the following lengths, viz. : 3, 3.8, 4.7, 5.3, 6.5, 8.3, 10.5, and 12 inches. Arrange them side by side on a table, a convenient distance apart; commencing with the longest, raise and let drop upon the table successively each bar, and determine by the pitch of the noise (every noise has pitch) produced on striking the table the relation between the vibration period of bars and their lengths.

EXP. 163. Provide three pine bars, each 4 inches long and 1 inch wide, and respectively $\frac{3}{16}$, $\frac{6}{16}$, and $\frac{3}{4}$ inch thick. Drop them, and determine the relation between the vibration period of bars and their thickness.

VIBRATIONS OF BARS FIXED AT ONE END.

EXP. 164. Take a lath about 1^m long and 3 to 5^{cm} wide, and about 5^{mm} thick, lay it upon a table with about two-thirds its length projecting beyond the table, press it firmly upon the table with one hand, about 6^{cm} from the edge of the table, and with the other hand bend the free part upward and let go. Notice the pitch of the vibration. Then shorten the projecting part one-fourth, one-half, three-fourths, etc., at a time, and notice the change in pitch.

LONGITUDINAL VIBRATIONS.

EXP. 165. Grasp between the thumb and fore-finger of one hand a glass tube (about 1^m long and 6^{mm} in diameter) about one-third its length from one end. Place on the palm of the

other hand a damp cotton cloth, and grasp the tube just below
the thumb and finger and quickly draw it along the tube (the
longer part) to the end. Repeat this movement several times
until you obtain loud, shrill sounds as the result of *longitudinal
vibrations.*

Exp. 166. Take a tube half the length of the last and produce longitudinal vibrations. The time of a complete vibration
being that required for the sonorous pulse to run twice to and
fro over the tube, how should the pitch of the latter compare
with that of the former?

Exp. 167. Grasp in a vice, in a nearly horizontal position, a
steel rod of about the same dimensions as the first glass rod,
and about one decimeter from one end. Sprinkle powdered rosin
on a leather glove, e.g., a dogskin or kid glove, and with the
rosined fingers excite vibrations by friction lengthwise the rod.
Suspend by a string a small ivory ball so that it may touch the
end of the rod nearest the vice. The ball will bound back and
forth against the rod when it is excited, thus showing that the
vibrations are longitudinal.

Exp. 168. Ascertain the wave-length in accordance with the
above in wires of different substances by throwing them into
longitudinal vibrations, and bringing them into unison (by varying the length) with that of a tuning-fork whose vibration rate
is known. Then calculate the velocity of sound in each substance.

DIAPASONS.

Exp. 169. Take a diapason (a large tuning-fork), and by means
of sealing-wax attach to the end of one of its prongs a narrow,
thin piece of sheet brass, cut at one end to a fine point, or a
very fine (No. 30) wire. Prepare a smoked glass by passing
one of its surfaces over a candle flame. Set the fork in vibration, and draw the style over the smoked surface with uniform
velocity, allowing the style lightly to touch the glass. Next draw
a bow rapidly and with slight pressure at a point about two-

thirds the distance from the end of the prong to the handle, and again draw the style across the smoked glass. This should give a graphical representation of the harmonic of the fundamental of the fork. If you do not succeed in this, bow the fork as before, and lightly touch with the extremity of the finger-nail a point about one-third the length of the prong from its extremity.

The sound of a tuning-fork, when set in vibration in the usual manner, contains, beside the fundamental, numerous overtones ; but the interval between them and the fundamental is very great. Consequently, the overtones are very evanescent, and soon leave the fundamental practically pure. This important acoustical property is very much increased when the stem is applied to a table or resonance-box, which reënforces the fundamental at the expense of the others.

Exp. 170. Produce the fundamental note, and, while it is sounding, draw the bow so as to give the harmonic, and immediately apply the style to the smoked glass. A curve should thus be obtained, resulting from the two systems of vibrations, consisting of smaller and shorter wavers superimposed upon larger and longer ones. If you succeed in obtaining good results, and wish to preserve them, first hold the blackened surface in the vapor of boiling alcohol, to remove the grease. Then pour amber varnish over it, as when varnishing a photographic negative, and allow it to dry.

Exp. 171. Take a glass plate of the same size as a stereopticon slide. Blacken it as in the foregoing experiments, but do not apply a very thick deposit of soot, so as to make it quite opaque. Draw the style of the fork when not vibrating lengthwise the blackened surface, and about one-third its width from the edge. A simple straight line will be the result. Produce the fundamental, and draw the style across the plate parallel with the line previously drawn. Produce the harmonic alone if possible, and draw the style across the plate parallel with the last two lines. Finally, produce the fundamental and the harmonic together, and once more draw the style across the plate. Varnish the plate, and preserve it for use in projecting with the porte-lumière graphical representations of actual vibrations.

Exp. 172. Set a tuning-fork in vibration, and touch that string of a violin which is nearest its own pitch, and move it along the string to or from the bridge until a length of string is obtained which will vibrate in sympathy with the fork, when a loud sound will be given forth by the string.

SAVART'S BELL, CHLADNI'S PLATE, AND INTERFERENCE-TUBE.

Exp. 173. Set in vibration, by bowing, a Savart's bell (see catalogue of apparatus to illustrate sound), and notice the wavy sound produced as the result of interference of the fundamental with its overtones. The fundamental is produced by the division of the edge into four segments.

Exp. 174. By repeated trials determine the fundamental pitch of the bell. Produce the fundamental, and, at 90° from the point bowed, hold as near the edge of the bell as convenient without touching, one end of the open resonance-tube. The sound is strongly reënforced. Place the plunger in the remote end so as to convert it into a closed tube, preserving the same distance from the bell, and repeat; the sound is only feebly reënforced. Gradually move the plunger inward until the maximum reënforcement is obtained, and determine what part the length of an open tube a closed tube should be to reënforce a sound of a given pitch.

Exp. 175. With the same bell produce a sound of higher pitch than the fundamental, and, with the plunger, adapt the length of the closed tube to the sound.

Exp. 176. Produce the fundamental, and reënforce with the closed tube its lowest overtone.

Exp. 177. Remove the bell from its iron support, and mount in its place the Chladni's plate. Scatter evenly over its surface fine writing-sand from a wooden sand-box, such as furnished by stationers. Bow the center of one edge with a rosined bow, producing the fundamental of the plate, known by the sand arranging itself in diagonal lines, so as to divide the plate into four segments. Note that in whatever way the plate is divided, there is always an *even* number of segments formed.

EXP. 178. (*a*) Bow the plate near one of its corners. (*b*) Bow the plate in the middle of the edge next you, and, with the thumb and one finger of the left hand, damp the left edge at points one-fourth the length of the edge from each of its extremities. (*c*) Obtain a variety of figures by varying the points bowed and the points damped. Make drawings of each, note the pitch of each, and describe the method by which each was produced.

EXP. 179. Produce the fundamental by bowing the center of one edge, hold the orifices of the two prongs of the interference-tube over the *opposite* segments, and adjust the length of the tube so as to strongly reënforce the sound. Then place the orifices over adjacent segments, and at the *same distance from the plate as before.* Interference causes a destruction of sound.

EXP. 180. Produce "hydrogen tones" by introducing a flame of illuminating gas (at the narrow orifice of a glass tube) about 2.5^{cm} long into a glass tube about 0.8^m long and 8^{mm} bore. At a little distance from the flame rotate the mirror furnished for manometric-flame experiments. Obtain tones of different pitch, and note the appearance of the vibrating flame in the rotating mirror.

EXP. 181. Procure a tin flageolet at a toy shop. Over the mouth of the instrument slip a rubber tube, and connect the other end of the tube with a gas-tube nipple or other source of condensed air or gas of any kind. Obtain a glass tube from 10 to 15 ft. long, with a bore large enough to receive nearly the whole of the tapering flute. Turn on the gas slowly; the first sound heard will probably be the lowest sound produced by the flute reënforced by the glass tube; on turning on the gas with greater force, higher notes break forth one after another. As

many as twenty distinct notes have been produced in this way.
If precipitated silica is scattered along the bore of the tube, it
will collect at the nodes of the tube corresponding to each
tone.

EXP. 182. Allow the corner of a card to tap the edges of the
siren plate as it is rotated, and you obtain all the phenomena
derivable from the expensive Savart's wheel.

For other experiments in the phenomena of sound, see the
admirable little volume on Sound in the Experimental Science
Series, by Prof. A. M. Mayer.

LIGHT.

RADIOMETER.

EXP. 183. Light a match, and hold the flame three or four inches from a radiometer.

EXP. 184. Place a radiometer three or four feet from a gas-flame.

EXP. 185. Hold the palm of your hand about an inch from a radiometer.

EXP. 186. Find the greatest distance from lights of different kinds, e.g., candle, kerosene, spirit, and Bunsen (both the light-giving and colorless) flames, and the electric light, that motion in the radiometer can be produced, and thus compare the mechanical power of the ether waves proceeding from the different sources.

DAYLIGHT PHOTOMETRY.

EXP. 187. (PICKERING.) Provide a box about 6 ft. long, 1 ft. wide, and 1½ ft. deep. A wooden frame covered with black paper will answer the purpose. Cut a circular hole about 4 in. in diameter in one end, and cover it with blue glazed paper with the white side out. Let drop a drop of melted sperm-candle wax upon the center of the disk, rubbing it around with the finger so as to cover a circular space of about the size of a silver dollar. Take a lath a little longer than the box, and about two inches wide, and cut a small hole in the end of the box containing the paper disk large enough to allow the lath to pass through it into the box, leaving one end projecting from the box. Upon the end of the lath in the box mount a lighted candle. The box should be properly ventilated by holes, so that the candle may not become extinguished. Taking hold of the exposed end of the lath, push the candle so far away from the paper disk that, in a room where the intensity of light is to be measured, the paper disk will appear dark in its center. Then draw the candle slowly forward until the center appears neither darker nor brighter than the center of the disk. Measure the distance of the candle from the disk. (If the lath has a scale of inches marked off on it, the distance will be very easily ascertained by observing the portion of the lath extending from the box, and deducting that from the length of the lath.) Unity divided by the square of this distance gives a measure of the comparative brightness of the daylight under various circumstances. Carry the box into different parts of the same room, and into different rooms, and measure the intensity of the light.

Exp. 188. With the daylight photometer measure the fading of the light at twilight. Other interesting experiments may be made with the same apparatus, by observing the intensity of light during an eclipse, and by comparing moonlight, and light of the Aurora, with daylight.

MEASUREMENT OF REFRACTION.

Exp. 189. Take "a tank with platinum electrodes," such as furnished by the author for projecting "phenomena attending electrolysis." Pour water into the tank, leaving about a quarter of an inch of electrodes exposed. Prepare two paper scales long enough to extend across the glass between the brass frame-work, and 4^{mm} wide, making the divisions of the scale 2^{mm} each. With flour paste apply one of these scales to the external surface of one of the glass sides, parallel with and about 1^{cm} below the surface of the water. Apply the other scale above and parallel with this, and so that the lower edge of the paper will be on a level with the upper extremities of the platinum electrodes. Place the eye about on a level with the surface of the water, so that it can, without moving, see both scales. If the eye is placed directly opposite one of the electrodes, it will appear at the same point on both scales; but the other electrode will appear at different points on the two scales, and the difference will be the amount of refraction. Move the eye to different positions, and observe the amount of refraction in each. It will be still better to set up vertical wires outside the tank and view them through the liquid.

LENSES.

The lens accompanying the porte-lumière is admirably adapted to the following experiments.

EXP. 190. *To find the focal length of a convex lens.* Hold a convex lens in the sunlight, its face at right angles to the sun's rays. Behind the lens and parallel with it hold a piece of light brown paper. Move the paper back and forth until the circle of light projected upon it is smallest and brightest. The distance of the paper from the center of the lens is its focal length.

EXP. 191. In a darkened room hold a candle-flame a few feet in front of the convex lens, and behind the lens a screen of white paper or white cotton cloth. Move the screen back and forth until a distinct inverted image of the flame is formed on the screen. Measure the respective distances of the image and object from the lens; also obtain as nearly as possible corresponding dimensions of the image and object, and verify the following formula : —

$$\frac{O}{I} = \frac{d}{d'},$$

in which O represents a given dimension of the object, and I a corresponding dimension of the image, and d and d' the respective distances of the object and image from the lens.

EXP. 192. Place a convex lens facing a window, and at a considerable distance from it. Upon a screen behind the lens project a distinct image of the window-frame. Observe that the distance of the image from the lens is greater than its focal length. Carry the lens nearer to the window; the distance of the image from the lens increases. Verify the following formula : —

$$\frac{1}{o} + \frac{1}{i} = \frac{1}{f},$$

in which o and i represent the respective distances of the object and image from the lens, and f its focal distance.

EXP. 193. Hold a convex lens at a considerable distance from the window, as before ; ascertain the distance of the lens from the window and from the image on the screen, and calculate its focal length from the last formula given.

EXP. 194. Look with one eye through a bi-convex lens at a piece of engineer's paper (divided into small squares) placed at its focus ; with the other eye look at a piece of the same paper placed at a distance of ten inches, and determine how many squares seen with the naked eye are contained in one seen through the lens, and thus judge approximately of the magnifying power of the lens.

EXP. 195. Repeat the last experiment, using, in place of the lens, a visiting-card pierced by a pin-hole.

THE RAINBOW.

EXP. 196. Fill a glass bulb about $1\frac{1}{2}$ in. in diameter (those furnished for air-thermometers answer the purpose) with a filtered solution of common salt in water. Cover the aperture of the porte-lumière with a black cardboard, so as to completely exclude the light from a darkened room. Cut a hole in the center of the cardboard of the same diameter as the bulb, and allow a circular beam of light to pass through it and also through a hole of about 4 in. diameter in the center of a white cardboard about 2 ft. square, and strike the bulb placed at a distance of about 2 ft. in front of the white cardboard. A miniature rainbow will be reflected back from the bulb upon the screen around its aperture. Any spot on the screen where red, for instance, appears, means that an eye situated at that point would see red in the glass bulb. Every other color, unless the eye was moved, would require another bulb in the proper relative position.

AFTER-IMAGES.

EXP. 197. Admit light into a darkened room by means of a porte-lumière through a slit about 1^{mm} wide, and project a bright spectrum upon a screen. Let the spectators look steadily at the spectrum for five or ten minutes ; then suddenly cut off the light, and, at the same instant, turn on a gas light, and a reversed spectrum will be seen on the screen, *i.e.*, the complementary of the former colors will be seen as an after-image.

EXP. 198. Prepare a circular disk of tin 10^{cm} in diameter ; punch holes about 4^{mm} in diameter equal distances (say from 3 to 4^{cm} apart) from one another in a circle within about 5^{mm} of the edge of the disk. Cut a hole in the center of the disk about 7^{mm} diameter, and place it on the spindle of the rotating apparatus. Place the disk in the path of a beam of light introduced into a dark room by a porte-lumière, and beyond the disk a lens, and focus the holes on a screen. Rotate slowly : a flickering light will be produced on the screen. Rotate rapidly, and the light will appear steady, due to *persistence of vision*.

EFFECT OF CONTRAST.

EXP. 199. On a sheet of white paper place an opaque ball, *e.g.*, a base-ball. Darken a room, and admit a small quantity of indirect sunlight through a space in one window about 1^{dm} wide. Place the paper and ball so that a shadow of the ball will be cast upon the paper. On the same side of the ball that the shadow is place a kerosene flame, at a short distance, so as to cast a shadow on the opposite side. So regulate the position of the paper and ball that the two shadows will have about the same depth. The shadow cast by sunlight will be yellow ; that cast by the flame, blue. Explain.

EXP. 200. Obtain a strip of cardboard about 40^{cm} × 6^{cm}, and a pan of vermilion water-color pigment. With a camel's-hair brush, by repeated washes, paint a portion of the strip at one end about 6^{cm} wide very deep, so that it will appear quite dark. Then

gradually grade the depth of the color up from this end toward
the other end, leaving a portion about 6^{cm} wide at one end un-
colored. Let the grading be so neatly and carefully done that the
eye will not detect the lines of separation of the different grades.
This may be effected partly by varying the number of washes,
and partly by thinning the washes with water. The effect is
pleasing to the eye, and the phenomenon is instructive, as the
effect of different depths of the same color are plainly depicted.

Now with the same pigment paint a piece of plain white paper
of uniform depth throughout, and about the same as the inter-
mediate depth on the cardboard. Cut out of this paper circles
about 15^{mm} in diameter, and paste them about 5^{cm} apart, cen-
trally and lengthwise across the cardboard. Although all the
circles have the same depth of color, they will appear, as the
effect of contrast, to be of very different depths. To make
the deception apparent, it is only necessary to take another
strip of cardboard (or paper) of the same size as the first, cut
holes about 13^{mm} in diameter, to correspond with the circles, and
lay it over the first cardboard so as to conceal all but the circles,
when the latter will all appear to be of uniform depth.

Exp. 201. Get papers of as great variety of colors as possible,
and cut out of each squares of 6^{cm} edge. Also cut circles of
1^{cm} diameter out of the same colors, two of a kind. Place two
circles of the same color upon squares of different colors, and it
will be difficult to persuade yourself that the circles have the
same color until you remove the circles from the colored squares
and place them side by side. By many experiments verify the
following

RULE :

*If we surround one color with another color, the former will
apparently be changed, as if some of the complementary of the
latter had been mixed with it.   Or, when any color of the chromatic circle (Fig. 282, Physics) is brought into competition or
contrast with any other color, the former is driven farther from
the latter in the circle.*

In case two colors are brought into juxtaposition and there is not great inequality in their areas, the two colors mutually drive each other apart. This may be shown by placing the squares in juxtaposition. Also slip one square behind the other so as nearly to conceal one by the other. In all these experiments it is best to stand at some distance from the objects under examination.

Exp. 202. Introduce into the porte-lumière in place of a slide a green glass, having any design cut out of opaque paper pasted on it, and a black design on a green ground will appear on the screen; but on bringing another light into the room or turning up the gas, the black design will at once appear as a brilliant pink. A glass of any other color may be used, and a design of its complementary color will appear.

Exp. 203. Paint figures on a white cardboard with chrome yellow. Illuminate the card in a darkened room with a salted Bunsen flame. The figures nearly disappear. Explain.

Exp. 204. Pour a little blue coloring solution into a glass beaker or large test-tube, and place behind it a black cloth; the larger portion of color disappears. A white cloth or white paper brings out the color more strongly. Explain.

Exp. 205. Lay a piece of gold leaf on a piece of glass, and look through it at the sun. Explain the change in the color of the gold leaf.

POLARISCOPE.

Exp. 206. Remove the analyzer A, Fig. 293, Physics, from the polarizer, and examine light reflected from the top of a varnished table. Rotate the analyzer, and see whether the light appears equally bright in all positions. Observe whether the color of the wood and its grain is seen better in some positions than in others.

Exp. 207. Place a coin under several plates of glass, and allow a strong light to fall upon it. Examine the reflected light with an analyzer, and see whether in those positions in which the light is polarized the coin is visible.

COLOR-BLINDNESS.

Exp. 208. Fill one of the tanks accompanying the porte-lumière with a solution of sulphate of copper, and look through it at various colored objects, and you will get an approximate idea of the appearance of things to a red-blind person. Or,

Exp. 209. Darken a room, and with a porte-lumière introduce a beam of light, causing it to pass through the tank, and afterwards through the convex lens. Let the light fall upon colored objects. Colored glasses may be used. When a yellow glass is used, the condition of the spectators is analogous to that of violet-blind persons, or of those who examine colors by gaslight, which is deficient in violet.

THE TELESCOPE.

By reference to Figs. 269 and 295 of the Physics, it will be seen that the distance of the image (ab) formed by a convex lens (*e.g.*, the object-glass) from the lens is greater for near objects than for objects farther off. Hence the eye-glass must be slightly farther from the object-glass in viewing near objects than in viewing objects farther away.

It will be seen that the distance of the image from the object-glass is somewhat greater than its focal length. On the other hand, the eye-glass must be brought somewhat nearer the image than its focal length. Hence the distance between the two lenses is *nearly* the *sum* of the focal lengths of the lenses.

Exp. 210. Place the lens accompanying the porte-lumière (or a lens having a focal length of 8 to 12 inches) from 4^m to 8^m from a window. Mount a white cardboard, and project upon it a distinct image of the window-frame. On the side of the cardboard opposite the image write a word with pen and ink. Take a convex lens of two inches to four inches focal length, and locate it so that the writing on the card can be seen through it distinctly and much magnified. Now withdraw the card, and the image of the window-frame will be seen much magnified.

In a darkened room a candle (or better, a gas or kerosene) flame may be substituted for the window-frame.

Use eye-glasses of different powers.

Prof. Crawford suggests the use of the projecting lens of a lantern for the object-glass, and for the eye-glass a simple hand magnifying glass, and adds, that "he finds that as a rule students have most remarkably vague notions as to the functions of a telescope, and such an experiment," as indicated above, "helps to clear up their ideas."

EXP. 211. Provide a white card 10^{cm} long and 6^{cm} wide, rule it with ink widthwise with heavy lines 1^{cm} apart, number the lines from bottom upward with plain figures, and place it in a vertical position on a black background with lines horizontal. At a suitable distance, say from three to eight rods, look at the card with the right eye through a telescope, focusing it so that the lines and figures can be distinctly seen. Then look at the card with the naked left eye, bring the image seen with the naked eye and the magnified image so that the adjacent vertical edges touch, and their lower edges coincide, measure the hight of the former on the latter by means of the numbered lines. The hight H of the larger image ($H = 10$) divided by the hight h of the smaller image will give the magnifying power m of the telescope ; that is

$$m = \frac{H}{h}.$$

EXP. 212. Ascertain the focal lengths F and f of the object-glass and eye-glass ; then

$$\text{the magnifying power } m' = \frac{F}{f}.$$

It will be found that m and m' are nearly equal in value. Why, if you use different eye-glasses (f) with the same object-glass (F), do you obtain different magnifying powers?

EXP. 213. Allow sky-light to pass through the telescope and form a distinct image of the object-glass on a white screen held at a suitable distance from the eye-glass. The diameter of the object-glass, divided by that of the image, is the magnifying power of the microscope.

Exp. 214. Determine by the first method described above the magnifying power of opera-glasses, looking at a card with one eye through one of the cones and at the same card with the other eye naked.

EXPERIMENTS WITH THE PORTE-LUMIÈRE.

The following experiments are intended to be supplementary to those contained in Dolbear's invaluable work on the "Art of Projection."

Exp. 215. Introduce a horizontal beam of light into the darkened room. Place a table so that its top will be from five to ten inches below the beam, and on the table in the path of the beam place the lens, and on the opposite wall or screen will be projected a large circular field of light. Place an ordinary stereopticon slide in the slide-holding disk. Move the lens back and forth until a distinctly-defined image is formed upon the screen. See Fig. 268, Physics.

Exp. 216. Introduce the disk with $\frac{1}{4}$ inch aperture. On the screen will appear several overlapping circles of light, illustrating in an interesting manner multiple reflection (Physics, p. 343).

Exp. 217. Cut small holes of triangular, square, and other shapes in pieces of cardboard, and cover the aperture of the porte-lumière. The image of the sun projected upon a distant screen in every case is round, i.e., it is independent of the shape of the hole. (See Deschanel's "Natural Philosophy," ¶ 683.) Now interpose the lens.

Exp. 218. Place the disk a (Fig. 13) with adjustable slit in position, and in the path of the beam and at a distance from the slit about equal to its focal length, the lens l, and at a distance of 2^m to 4^m from the lens, a screen s about 1^m square. Focus the slit upon the screen, and then interpose the bisulphide

of carbon prism p about 8^{cm} in front of the lens. and move the
screen so as to receive the spectrum, as at s', preserving the
same distance from the prism. Rotate the prism slowly on its
axis until at least 15 to 20 dark lines are
seen in the spectrum. More lines may be
obtained by using more prisms, so as to
increase the dispersion.

Caution. — The cement which holds the
glass sides on the metal frame of these
prisms is insoluble in bisulphide of carbon.
but is soluble in water : hence the latter
liquid should never be used in these prisms.

Exp. 219. Cover the disk of 1 inch aperture with coarse lace or punctured card
paper, such as is used for book-marks, and
project interference phenomena.

Exp. 220. Project the phenomenon illus-
trated in Fig. 257, Physics.

Fig. 13.

Exp. 221. Project the piece of glass used in the last experiment, and rotate it on its longest axis. In certain positions the
light will be intercepted by total reflection, and a deep shadow
cast upon the screen.

Exp. 222. Fill a small test-tube, about 6^{mm} in diameter, with
water, and project it upon the screen. Only a very narrow line
of light will succeed in passing through the tube. Fill one of
the tanks which accompany the porte-lumière with water, and
thrust the tube into the tank. The water in the tank will cor-
rect the refraction, except that produced by the glass tube.

Exp. 223. Thrust an empty tube into the water-tank.

Exp. 224. Thrust a tube filled with bisulphide of carbon into
the water-tank.

Exp. 225. Make a saturated solution of sulphate of zinc, and
half fill a tank with the solution. With a pipette carefully fill
the tank with pure water. Let drop one or two drops of bisulphide of carbon into the liquid. The last liquid should float

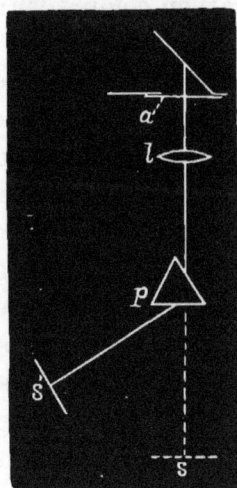

midway, partially immersed in the two liquids. Account for the many interesting phenomena.

Exp. 226. Repeat the last experiment, using bisulphide of carbon colored with iodine.

Exp. 227. Pass a beam of light through the lens ; strike together two blackboard brushes, just above the light, after it emerges from the lens, so as to render visible the cone of rays with its base on the lens, also the cone of rays whose vertex is at the focus and base on the screen. The smoke from touchpaper (see foot-note, p. 278, Physics) may be used to advantage.

Exp. 228. Repeat the last experiment, intercepting a portion of the light by using the circular 1 inch aperture.

Exp. 229. Introduce in front of this, or a smaller aperture, a concave lens, and show the divergence caused by the lens.

Exp. 230. Using a small aperture, introduce a small beam of light, and reflect it, by means of a mirror, into different parts of the room. Explain why the angle which the reflected beam makes with incident beam is double that which the incident beam makes with the mirror. Explain why the spot of light thrown upon the walls is usually much elongated in one direction.

Exp. 231. Move a white screen slowly back and forth through the foci of the different colors produced by the lens, and observe the effects of chromatic aberration. Observe that the red rays have their focus furthest from the lens.

Exp. 232. Mix lamp-black with French polish, and thin it with spirits of turpentine. With a camel's-hair brush apply a coating to a slip of glass. Apply other coatings as fast as it becomes dry until the glass becomes quite opaque. With the point of a penknife blade draw on the glass any designs or diagrams which you may desire to project.

Exp. 233. Remove the mirror of the porte-lumière, and introduce the disk with half inch aperture. Inverted images of the landscape toward which the window is directed will be formed on the walls of the room. Persons passing by the window, at suitable distance from it, will appear walking feet upward. Let persons from a distance run toward the window, and observe the effect on the size of the images.

EXP. 234. Cover the orifice of the porte-lumière (its mirror being removed) with the lens. Behind the lens move a piece of white cardboard slowly from the lens until a distinct inverted image of a distant house, tree, etc., is projected upon the card. These experiments should be performed at such time of the day as the surfaces turned toward the lens are illuminated by the direct rays of the sun.

EXP. 235. Pour dilute sulphuric acid into the "miniature battery," and project upon the screen. A volley of hydrogen bubbles will appear to fall (because the battery will necessarily appear inverted on the screen) from the zinc rod. Connect the electrodes, and instantly the larger portion of the bubbles will escape from the copper rod, a few still escaping from the zinc. None are seen to pass through the liquid from zinc to copper.

EXP. 236. Place in position the "movable slide" to show wave-motion; draw the slip with wavy edge past the disk with parallel slits, and long lines of light will play up and down on the screen, while the wave will traverse the screen horizontally.

EXP. 237. Take the two slips with wavy edges furnished with the above apparatus; place them so that crest will touch crest, and project upon the screen. A series of elliptical areas of light extend horizontally across the screen; the vertical diameters represent the intensity of the resultant of two similar sets of waves when they so interfere that the same phases of both coincide. Gradually move the upper slip horizontally, letting it rest all the time on the upper edge of the lower slip until the crest of the upper fills the trough of the lower slip, when all the light will be cut off from the screen. The gradual narrowing of the luminous areas, until the light is entirely extinguished, will show all the results of partial neutralization by interference until the opposite phases exactly concide, when there is a complete destruction.

EXP. 238. To show the toughness of liquid films by projection, arrange the porte-lumière for vertical projection, as described, pp. 40–44, Dolbear's "Art of Projection." Over the condenser rest a shallow transparent vessel containing a solution of saponine, one part of the powder to sixty of water. The depth of the liquid may be 4 or 5^{mm} or less.

Construct a wooden frame-work like Fig. 14; the length ab being such that it may enclose the glass vessel. Through the uprights bd and ac are holes e and f, through which a piece of glass tubing gh is passed, fitting so as to rotate with slight friction. The exterior diameter of this may be 4 or 5^{mm}. Around the middle m is wrapped a fibre of unspun silk, whose lower end is fastened to the middle of a piece of steel needle about 2^{cm} or 3^{cm} long. Magnetize this, and let it hang horizontally over the middle of the saponine solution close to the surface. Focus so that a distinct image of the needle is formed on the screen. Bring a bar-magnet near, from side to side, and observe that the needle oscillates freely in obedience to it. Now turn the glass gh till the needle rests on the surface of the liquid without being immersed. Repeat the use of the bar-magnet. The needle fails to oscillate, or oscillates but slightly, being held tightly by the surface film. Again depress the needle until wholly immersed. Repeat the use of the bar-magnet. The needle oscillates freely, though slightly impeded by the viscosity of the liquid. Only at the surface is it held approximately immovable.

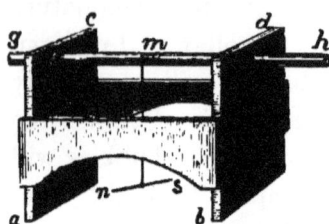

Fig. 14.

NOTE. The author is prepared to furnish all appurtenances necessary for vertical projection, including a tank with a plate-glass bottom.

LAWS RELATING TO ELECTRIC CURRENTS.

I. *To produce a current there must be two points at different potentials separated from each other by a resisting medium.*

To produce a continuous current these points must be maintained at different potentials. A current is an apparent transference of electricity from one point to another to produce equalization of potential.

II. *The difference in potentials between different points of a circuit varies as the resistances between the respective points.*

It is found experimentally, by measuring with a delicate electrometer, that, between any two cross-sections A and B of a homogeneous wire, in which a uniform current is kept flowing, there exists a difference of potentials, and that if the wire be of uniform section throughout, the difference of potentials is in direct proportion to the length of the wire between the cross-sections.

Suppose that A, B, and C represent consecutive points in a circuit, d the difference in potentials between A and B, and r the resistance between the same points, d' the difference in potentials between points B and C, and r' an unknown resistance between the same points, then

$$d : d' :: r : r', \text{ whence } r' = \frac{d'r}{d}.$$

"The potential of a point" is the difference between the potential of that point and that of the earth reckoned as a zero.

III. Ohm's Law. *The strength of a current is directly proportional to E.M.F., and inversely proportional to resistance.* This relation is generally expressed

$$C = \frac{E}{R}, \text{ or } C = \frac{E}{r + R},$$

when it is important to separate the internal from the external resistance. It will be seen that the foregoing law is only an application to a specific case of a universal fact; viz., that an effect is directly proportional to that which tends to produce it, and inversely proportional to that which tends to oppose it.

IV. *Resistance varies* (1) *as the length of conductors; it varies* (2) *inversely as the areas of cross sections of conductors or the squares of diameters of cylindrical conductors; it varies* (3) *inversely as the specific resistances of the substances used for conductors.*

V. *E.M.F. depends upon the nature and condition of the substances which compose the battery, and is independent of the size of the plates and their distances apart.*

VI. *Where there are no leakages the strength of a current is equal at all points in a circuit.*

GROUPING OF CELLS.

VII. *For a given battery of cells the most effective way of grouping them, when they are required to work through a given external resistance, is that in which the difference between the external and internal resistance is least. Hence a given battery works to the greatest advantage when the external and internal resistances are equal.*

It must not be inferred from the latter statement that when the internal resistance is less than the external resistance, that we should increase the internal resistance for the sake of the resistance. The internal resistance of itself is a positive disadvantage. The necessity for so great an internal resistance results only from the fact that it is an unavoidable accompaniment of high E.M.F. when that is obtained by connecting in series the requisite number of cells.

To get the maximum current through a high resistance it is clear that we will gain more by the increase of E.M.F. resulting from connecting cells in series, than we will lose by the accompanying increase of resistance, since the latter is but a small part of the whole. If the external resistance is very small so that the resistance of the cells is a large part of the whole, we shall gain very little by arranging them in series, since we should increase the total resistance of the circuit in nearly the same ratio as we should the E.M.F. If we arrange them abreast, we shall neither gain nor lose in E.M.F., but we shall reduce rapidly the total resistance of the circuit.

VIII. *The current will be a maximum when the number of cells connected abreast is numerically equal to* $\sqrt{nr \div R}$.

Thus, suppose the number of cells is 40, and $r = 3$ ohms, and $R = 8$ ohms, then $\sqrt{40 \times 3 \div 8} = 3 +$, nearly 4; hence, four cells should be connected abreast, and the ten groups connected tandem.

The following law as given by Gordon will be found very convenient.

To obtain a maximum current, the ratio of the number of cells in series to the number of cells connected abreast should equal the ratio of the external resistance to the resistance of a single cell. That is, C is a maximum when $\dfrac{N}{n} = \dfrac{R}{r}$, or when $R = \dfrac{N}{n}r$, in which N is the number of cells in series and n the number connected abreast.

LAW OF DIVIDED CIRCUITS.

IX. *When a circuit is so constituted that at some point there are two paths open to the circuit, the current will divide between these two paths in the inverse ratio of their resistances, and the joint resistance of the two paths will be neither the sum nor the difference of their respective resistances, but will be expressed by the product of these resistances divided by their sum.*

ESTIMATING WORK DONE BY A CURRENT.

X. *The mechanical work of a current may be calculated from the following formula, in which C is the current strength, and V is the difference in potential of the terminals of the instrument in which the work is done:—*

$$\frac{C \text{ ampères} \times V \text{ volts}}{745} = rate \text{ of doing work in horse-powers.}$$

For example, to find the rate at which actual work is consumed in an electric lamp, measure the whole current in ampères; measure the difference of potential (with an electrometer) between the terminals of the lamp in volts; multiply them together, and divide by 745; the result will be the number of horse-powers used up in the lamp. That is, a current of C ampères falling V volts will perform, in passing through the instrument, work at the rate of $x\left(= \dfrac{C \text{ ampères} \times V \text{ volts}}{745}\right)$ horse-powers.

XI. *The amount of chemical decomposition produced by a current in a given time varies as the strength of the current.* On this principle is constructed the voltameter, which measures the strength of the current by the amount of chemical action it effects.

XII. *The weight in grams of an element deposited by electrolysis is found by multiplying its electro-chemical equivalent (i.e., the atomic weight divided by the valency) by the strength of the current in ampères, and this product by the time in seconds during which the current electrolyzes.*

A current of one ampère strength will decompose about three grams of water per hour. A current of n ampères will decompose 0.0000973 n grams of water per second.

XIII. *The number of units of heat developed in a conductor is proportional, (1) to its resistance; (2) to the square of the strength of the current; and (3) to the time the current acts.*

A current of one ampère flowing through a resistance of one ohm, develops therein 0.00024 calorie of heat per second. Hence,

H (calories) $= C^2$ (ampères) $\times R$ (ohms) $\times t$ (seconds) $\times 0.00024$.

That portion of a current not expended in external work is "frittered down into heat," either in the battery or in some part of the circuit, or in both.

The absolute amount of heat generated by the oxidation of a given quantity of zinc is perfectly constant; but it may be distributed in various proportions between the battery and the external circuit.

Whenever the current heats a wire, produces decomposition, or performs work of any kind, each of these acts is accomplished at the expense of the heat in the battery. If the current turn an electro-magnetic engine which pumps water, or lifts a hammer, the battery loses heat by these mechanical acts. The precise amount of heat thus lost is restored by the falling of the water, or of the hammer. So also when water is decomposed in a voltameter, the battery loses an amount of heat equal to that which would be produced by the recombination of the separated oxygen and hydrogen.

Fauré corroborates these results in a series of careful experiments, in which he interposed increased resistance by increasing the length of the wire connecting the two ends of the battery. His main results are given in the following table : —

Length of Wire.	Internal Heat of Battery. (Units.)	Heat outside the Battery. (Units.)	Total. (Units.)
25	13127	4965	18092
50	11690	6557	18247
100	10439	7746	18185
200	8992	9030	18022

LAWS OF THE ELECTRO-MAGNET.

I. *In order to produce the most advantageous effect, the resistance of the helix of an electro-magnet should be equal to that of the portion of the circuit not included in the helix, i.e.*, to the rest of the circuit. *When several electro-magnets are used in the same circuit, the sum of the resistances of all the helices should be equal to the resistance of the rest of the circuit.*

Caution. Having decided correctly the resistance an electro-magnet should have, it is possible to make a serious error in selecting wire either too fine or too coarse. For example, suppose that an electro-magnet of 4 ohms resistance is to be made. If No. 20 wire is used, it will require about 170 yds. ; but if No. 32 wire is used, it will require only 16 yds. The latter would give only a few convolutions, which might not produce the maximum magnetic effect ; while the former might be so coarse that the outside layers would have little effect on the core ; hence the maximum effect in neither case would be obtained. In electro-magnets of ordinary size the best distance for the outside layer from the core is between three-eighths and half an inch. Or, in general,

II. *The thickness of the helix should be equal to the diameter of the core.*

It is apparent from the above laws that we would choose for a short circuit, or for a circuit where there is little other resistance, an electro-magnet of small resistance, *i.e.*, one made of coarse wire. On the other hand, for a long circuit, or a circuit of high resistance, an electro-magnet of high resistance, *i.e.*, one made of fine wire, should be ordered. Not because high resistance of itself is advantageous (it is a positive disadvantage), but in order to make the most of the existing current, weakened by the other high resistance in the circuit. we require

many turns of wire, the resistance which it brings with it being
a· necessary but an unwelcome adjunct. "The condensed reason
why we use fine wire — and a great deal of it — for circuits of
high resistance, is that the high resistance of the circuit greatly
enfeebles the current, and we must use fine wire to make the
best of the remaining strength of the current by a greatly-
increased number of convolutions."

Why is not a magnet containing many convolutions of fine
wire — in other words, a high resistance magnet — as efficacious
in producing magnetic effects when in a circuit of low resistance
as when in a circuit of high resistance? It would be equally·
efficacious if, by introducing the high resistance magnet, the
whole resistance of the circuit in both cases were proportionately increased, and thereby the current proportionately de-
creased. Take an example. Suppose the whole resistance of
a circuit including that of the battery is 20 ohms. Then a
magnet suited to this circuit should have a resistance of 20
ohms. The introduction of this magnet will double the resistance of the circuit and reduce the current strength one-half.
But if the entire resistance of the circuit is 1 ohm, and a magnet
of 20 ohms' resistance is introduced, the entire resistance of the
circuit will become 21 ohms, and the current is reduced to one
twenty-first its former strength. In the latter case, the advantage derived from the great number of convolutions of wire in
the magnet cannot compensate for the great reduction of
current.

Roughly, "the total length of the core, including both arms
and the back armature or connecting yoke, should be about
eleven times the diameter. When, however, the circuit is long
and the electric source is of feeble energy, the magnet should be
long and of small diameter. When, on the contrary, the circuit
is short and the current strong, the core should be of large
diameter."

"The attraction of magnets for prismatic armatures at a
distance is greatest when they are presented flatwise: but

when in contact, the attractive force is greatest when they are presented edgewise." — FISKE.

Electro-magnets with short cores charge and discharge more rapidly than those with long ones. Advantage is taken of this in constructing telegraph sounders and relays.

III. *The attractive force exerted by an electro-magnet is proportional to the diameter of the core and to the square root of its length.*

IV. *The attractive force of electro-magnets is proportional to the square of the strength of current for a like number of convolutions, and to the square of the number of convolutions for like strength of current.*

If the strength of the current (acting on the electro-magnet) and the number of convolutions in the helix vary at the same time, which is nearly always the case, since by increasing the number of convolutions without changing the battery we increase the resistance of the circuit, and thereby weaken the current,

V. *The attractive force of the electro-magnet is proportional to the square of the strength of the current multiplied by the square of the number of convolutions.*

VI. *The maximum of saturation depends solely upon the mass of iron contained in the electro-magnet irrespective of its form.*

VII. *The maximum degree of magnetization, of which a mass of soft iron is susceptible under the influence of the electric current, is more than five times as great as that which a corresponding mass of hardened steel is capable of retaining.*

ELECTRICAL UNITS ADOPTED IN PRACTICE.

Resistance. The resistance offered by 37.8^m (41.3 yds.) of pure copper wire, size No. 20 (New British standard Wire Gauge), diameter, 0.914^{mm} (0.36 in.), temperature 15° C., is 1 *ohm.* The *megohm* is 1,000,000 ohms.

The legal ohm is the resistance of a column of mercury 106^{cm} long and 1^{qmm} in section at 0° E.

Potential, electro-motive force. The difference of potential between the plates of a Daniell or Gravity cell, or the electro-motive force of one of these cells, is about 1 *volt.* The volt is the difference of potential or electro-motive force necessary to sustain a current of 1 ampère strength against a resistance of 1 ohm.

Current strength. A current flowing in a wire of resistance 1 ohm, between the two ends of which a difference of potential of 1 volt is maintained, is the unit of current, or current strength, and is called an *ampère.* It is a current of 1 coulomb per second.

Quantity of electricity. The amount of electricity conveyed in 1 second by a current of 1 ampère is called 1 *coulomb.* It is the quantity that would charge a condenser of 1 farad capacity, under an E.M.F. of 1 volt.

Electrostatic capacity. The capacity of a condenser, which would contain 1 coulomb of electricity when charged by an electro-motive force of 1 volt, is 1 farad. A *microfarad* is a millionth part of a farad.

Rate of doing work. The rate at which work must be done in maintaining a given current against a given resistance is measured by *EC*, in which *E* is reckoned in volts and *C* in ampères. Hence, rate of doing work, the power of a current, or the rate at which work is given out in a circuit, is measured by its *EC*. The unit employed is called a *watt*, or *volt-ampère*, and is the power developed by 1 ampère falling 1 volt.

Work done. The work done in 1 second, when the rate of working is 1 watt, or the work obtained by letting down 1 coulomb of electricity through a difference of potentials of 1 volt is 1 *joule*, or *volt-coulomb*.

EQUIVALENTS OF WORK.

$$1 \text{ joule or volt-coulomb} = \begin{cases} 0.101937^{kgm}. \\ 0.737324 \text{ ft. lb.} \\ 0.00024067 \text{ calorie of heat.} \\ 10^7 \text{ ergs.} \end{cases}$$

$1^{kgm} = 9.81$ joules ; 1 ft. lb. $= 1.35626$ joules.

EQUIVALENTS OF POWER.

$$1 \text{ watt or volt-ampère} = \begin{cases} 0.00134059 \text{ horse-power.} \\ 0.101937^{kgm} \text{ per second.} \\ 6.11622^{kgm} \text{ per minute.} \\ 0.000240670 \text{ calorie per second.} \\ 0.144402 \text{ calorie per minute.} \\ 10^7 \text{ ergs per second.} \end{cases}$$

1 horse-power $= 745.941$ watts ; 1 ft. lb. per second $= 1.35626$ watts ; 1^{kgm} per second $= 9.81$ watts.

SOURCES OF ENERGY.

Primary Sources.

1. Primordial energy of chemical affinity.
2. Solar radiation.
3. Energy of the earth's rotation.
4. Internal heat of the earth.

Secondary Sources.

A. *Potential.*
{
1. Combustibles.
2. Food of animals.
3. Ordinary water-power.
4. Tidal water-power.
}

B. *Kinetic.*
{
5. Winds and ocean currents.
6. Hot springs and volcanoes.
}

CLASSIFICATION OF VARIOUS FORMS OF ENERGY.

I. MECHANICAL ENERGY.
{
A. *Visible kinetic energy; i.e.,* energy of a body in visible motion.
B. *Potential energy of visible arrangement; e.g.,* a stone elevated above the earth.
}

II. MOLECULAR ENERGY.
{
C. *Kinetic energy of electricity in motion.*
D. *Kinetic energy of radiant heat and light.*
E. *Kinetic energy of absorbed heat.*
F. *Molecular potential energy; e.g.,* the so-called " latent heat."
G. *Potential energy of electrical separation.*
H. *Potential energy of chemical separation; e.g.,* oxygen and hydrogen.
}

A + B + C + D, etc., = a constant quantity.

TABLES OF REFERENCE.

SPECIFIC GRAVITY OF VARIOUS SUBSTANCES.

	Specific gravity.			Specific gravity.
Acetic acid	1.060	Deal, Norway		0.689
Alcohol	0.792	Diamond		3.530
Aluminium, sheet	2.670	Earth	from	1.520
Antimony, cast	6.720		to	2.000
Ash, dry	0.690	Ebony		1.187
Ash, green	0.760	Elm		0.579
Asphalt	2.500	Elm, Canadian		0.725
Basalt	2.950	Ether		0.716
Beech, dry	0.690	Feldspar		2.600
Bell-metal	8.050	Fir, spruce		0.512
Birch	0.690	Firestone		1.800
Bismuth, cast	9.822	Glass, flint		3.000
Bisulphide of carbon	1.293	Glass, crown		2.520
Boxwood	1.280	Glass, common green		2.520
Brass, cast	8.400	Glass, plate		2.760
Brass, sheet	8.440	Gold		19.360
Brick, common { from	1.600	Gold, 20 carats		15.700
{ to	2.000	Granite		2.650
Carbon, gas	1.760	Gun-metal (10 cop., 1 tin)		8.561
Carbonic acid, liquid	0.830	Gutta-percha		0.966
Cedar, American	0.554	Gypsum		2.286
Cedar, Lebanon	0.486	Human body		0.890
Cedar, West Indian	0.748	Hydrochloric acid		1.200
Cedar, Indian	1.315	Ice at 32°		0.930
Chalk	2.330	Iron, cast, average		7.230
Chestnut	0.606	Iron, wrought, average		7.780
Clay	1.900	India-rubber		0.930
Coal, anthracite	1.600	Iodine		4.950
Coal, bituminous	1.270	Ivory		1.820
Cobalt	8.800	Lead, cast		11.360
Concrete, ordinary	1.900	Lead, sheet		11.400
Concrete, in cement	2.200	Lignum vitæ		1.333
Cork	0.240	Lime, quick		0.843
Copper, cast	8.607	Limestone		3.180
Copper, sheet	8.780	Logwood		0.913

SPECIFIC GRAVITY OF VARIOUS SUBSTANCES. — *Continued.*

	Specific gravity.		Specific gravity.
Magnesium	1.750	Platinum, average	21.531
Mahogany, Honduras	0.560	Plumbago	2.267
Mahogany, Nassau	0.668	Quartz	2.650
Mahogany, Spanish	0.852	Rock salt	2.257
Maple	0.675	Saltpeter	1.900
Marble	2.720	Sand, quartz	2.750
Mercury	13.596	Sand, river	1.880
Milk	1.032	Sand, fine	1.520
Mortar, average	1.700	Sand, coarse	1.610
Muriatic acid	1.200	Silver	10.474
Naphtha	8.470	Slate	2.880
Nitric acid	1.217	Sulphur, natural	2.033
Oak, American, red	0.850	Sulphuric acid	1.840
Oak, American, white, dry	0.779	Tallow	0.940
Oak, live, seasoned	1.068	Tar	1.010
Oak, live, green	1.260	Teakwood	0.806
Oil, linseed	0.940	Tile, average	1.830
Oil, olive	0.915	Tin, cast	7.290
Oil, turpentine	0.870	Water, 32°	0.999
Oil, whale	0.923	Water, 212°	0.958
Phosphorus	1.830	Water, distilled, 39°	1.000
Pine, red, dry	0.590	Water, sea	1.627
Pine, white, dry	0.554	White metal (Babbitt)	7.310
Pine, yellow, dry	0.461	Willow	0.400
Pine, pitch	0.660	Zinc, cast	7.000
Pitch	1.150		

GASES.

	Specific gravity.		Specific gravity.
Air, 32°	1.0000	Hydrogen	0.6930
Ammonia	0.5367	Marsh gas	0.5596
Carbonic acid	1.5290	Nitrogen	0.9714
Carbonic oxide	0.9670	Oxygen	1.1057
Coal gas	from 0.3400 to 0.6500	Sulphuretted hydrogen	1.1912
Chlorine	2.4600	Sulphurous acid	2.2474
Hydrochloric acid	1.2540	Vapor of water	0.6235

AVERAGE COMPOSITION OF ALLOYS.

	Standard English gold coins.	Standard English silver coins.	Brass.	Dutch metal.	Gun metal.	Speculum metal.	Medal bronze.	German silver.	Solder, ordinary.	Pewter.	Type metal.	Fusible metal.	Shot.
Gold . .	92	..	..	..	..	..	..	..	..	..	..	..	..
Silver . }	8	93	..	..	..	..	..	..	..	..	..	..	..
Copper)		7	69	75	85	60	93	55	..	..	..	..	..
Tin . .	..	..	..	3	15	30	7	..	50	80	3	25	..
Zinc . .	..	..	31	19	..	..	..	27	..	..	..	..	..
Lead . .	..	..	..	3	..	..	..	..	50	20	80	25	98
Antimony	..	..	..	..	..	..	..	..	..	..	17	..	..
Arsenic .	..	..	..	..	..	10	..	..	..	..	..	..	2
Bismuth .	..	..	..	..	..	..	..	..	..	..	..	50	..
Nickel .	..	..	..	..	..	..	..	18	..	..	..	..	..

FRIGORIFIC MIXTURES FOR THE ARTIFICIAL PRODUCTION OF COLD.

NOTE. To obtain the following results, the temperature of the ingredients must be reduced previously to the given temperature by some of the other mixtures. The last four mixtures yield the temperatures given whatever may be the previous temperature of the ingredients. The temperatures given are of the Fahrenheit scale.

	Proportional parts, by weight, in the mixture.											
Water	..	..	..	..	..	..	..	..	..	..	..	..
Sal ammoniac . .	16	1	..	..	..	..	..	..	..	1	5	..
Nitre	5	..	..	..	..	..	..	..	..	..	5	..
Common salt . . .	5	..	..	..	1	..	..	..	1	2	10	5
Nitrate of ammonia	..	1	5	6	..	..	..	..	..	..	..	5
Sulphate of soda . .	..	..	6	..	..	..	..	..	..	..	..	..
Carbonate of soda .	8	1	..	..	..	..	..	..	..	..	..	..
Phosphate of soda .	..	..	..	9	..	..	..	..	..	..	..	..
Potash	..	..	..	..	..	..	..	4	..	..	..	..
Muriate of lime . .	..	..	..	..	..	..	5	..	..	..	..	..
Snow or pounded ice	..	..	..	..	1	8	4	3	2	5	24	12
Diluted nitric acid .	..	..	4	4	..	..	..	..	..	..	..	..
" sulphuric acid	..	..	..	..	..	10	..	..	..	..	..	..
Temp. of ingredients	+50	+50	+50	+50	+32	−68	+32	+32	..	..	..	..
" of the mixture	+4	−7	−14	−21	0	−91	−40	−51	−5	−12	−18	−25
Cold produced . .	46°	57	64	71	32	23	72	83	..	..	..	..

TABLE OF LINEAR DILATATIONS OF SOLIDS. — STEWART.

Name of Substance.	Length at 100° C. of a rod whose length at 0° C. = 1.000000.	Name of Observer.
Glass (tube)	1.000776	Roy and Ramsden.
Copper	1.001722	Lavoisier and Laplace.
Brass	1.001867	" "
Iron, soft (forged)	1.001220	" "
Steel (untempered). . . .	1.001079	" "
" (tempered yellow) . .	1.001240	" "
Cast iron	1.001072	Daniell.
Lead	1.002848	Lavoisier and Laplace.
Tin	1.001767	Daniell.
Silver	1.001951	"
Gold (standard of Paris, not annealed)	1.001552	Lavoisier and Laplace.
Platinum	1.000884	Dulong and Petit.
Zinc	1.002976	Daniell.

RELATIVE THERMAL CONDUCTIVITY OF METALS.

Silver.	100.0	Steel	11.6
Copper	73.6	Lead	8.5
Gold	53.2	Platinum	8.4
Brass	23.6	Palladium	6.3
Tin	14.5	Bismuth.	1.8
Iron	11.9		

SPECIFIC HEATS.

Lead	0.0314	Tin	0.0562
Iron	0.1140	Ether at 17°	0.5160
Glass	0.1900	Alcohol at 17°	0.6150
Gold	0.0324	Quicksilver	0.0333
Copper	0.0951	Oil of turpentine at 17°. .	0.4260
Brass	0.0940	Water at 0°	1.0000
Platinum	0.0324	Water mean between 0° and 100° . . . . }	1.0050
Silver	0.0570		
Zinc	0.0955		

CROSS-SECTION OF ROUND WIRES, WITH RESISTANCE AND WEIGHT OF
PURE COPPER WIRES, ACCORDING TO THE BIRMINGHAM WIRE GAUGE.
—GRAY.

Temperature 15° C.

B.W.G.	Diameter.		Area of Cross-section.		Resistance.		Weight (density = 8.95).	
No.	Ins.	Cms.	Sq. Ins.	Sq. Cms.	Ohms per Yard.	Ohms per Meter.	Lbs. per Yard.	Grms. per Meter.
0000	.454	1.1530	.1620000	1.0444000	.000152	.000167	1.884000	934.7000
000	.425	1.0790	.1420000	.9150000	.000174	.000190	1.651000	819.1000
00	.380	.9650	.1130000	.7320000	.000217	.000238	1.320000	654.8000
0	.340	.8640	.0908000	.5860000	.000272	.000297	1.056000	524.2000
1	.300	.7620	.0707000	.4560000	.000349	.000382	.822000	408.1000
2	.284	.7210	.0633000	.4090000	.000389	.000425	.737000	365.8000
3	.259	.6580	.0527000	.3400000	.000468	.000512	.613000	304.2000
4	.238	.6050	.0445000	.2870000	.000554	.000606	.518000	256.9000
5	.220	.5590	.0380000	.2450000	.000649	.000709	.442000	219.5000
6	.203	.5160	.0324000	.2090000	.000762	.000833	.377000	186.9000
7	.180	.4570	.0254000	.1640000	.000969	.001060	.296000	146.9000
8	.165	.4190	.0214000	.1380000	.001150	.001260	.249000	123.5000
9	.148	.3760	.0172000	.1110000	.001430	.001570	.200000	99.3000
10	.134	.3400	.0141000	.0910000	.001750	.001910	.164000	81.4000
11	.120	.3050	.0113000	.0730000	.002180	.002380	.132000	65.5000
12	.109	.2770	.0093300	.0602000	.002640	.002890	.109000	53.9000
13	.095	.2410	.0070900	.0457000	.003480	.003800	.082500	40.9000
14	.083	.2110	.0054100	.0349000	.004560	.004980	.063000	31.2000
15	.072	.1830	.0040700	.0263000	.006060	.006620	.047400	23.5000
16	.065	.1650	.0033100	.0214000	.007430	.008130	.038600	19.2000
17	.058	.1470	.0026400	.0170000	.009330	.010200	.030700	15.3000
18	.049	.1240	.0018900	.0122000	.013100	.014300	.022000	10.9000
19	.042	.1070	.0013900	.0089400	.017800	.019600	.016100	8.0000
20	.035	.0889	.0009620	.0062100	.025600	.028000	.011200	5.5600
21	.032	.0813	.0008040	.0051900	.030700	.033500	.009360	4.6400
22	.028	.0711	.0006160	.0039700	.040000	.043800	.007160	3.5500
23	.025	.0635	.0004910	.0031700	.050200	.054900	.005710	2.8300
24	.022	.0559	.0003800	.0024500	.064900	.070900	.004420	2.1900
25	.020	.0508	.0003140	.0020300	.078600	.085800	.003670	1.8200
26	.018	.0457	.0002540	.0016400	.096900	.106000	.002960	1.4700
27	.016	.0406	.0002010	.0013000	.123000	.134000	.002340	1.1600
28	.014	.0356	.0001540	.0009930	.160000	.175000	.001790	.8890
29	.013	.0330	.0001330	.0008560	.186000	.203000	.001540	.7660
30	.012	.0305	.0001130	.0007320	.218000	.238000	.001320	.6530
31	.010	.0254	.0000785	.0005070	.314000	.343000	.000915	.4540
32	.009	.0229	.0000636	.0004100	.388000	.424000	.000746	.3670
33	.008	.0203	.0000503	.0003240	.491000	.536000	.000585	.2900
34	.007	.0178	.0000385	.0002480	.641000	.701000	.000442	.2200
35	.005	.0127	.0000196	.0001270	1.260000	1.370000	.000229	.1130
36	.004	.0102	.0000126	.0000811	1.960000	2.150000	.000146	.0726

THE FOLLOWING TABLE EXHIBITS THE DECLINATION OF THE NEEDLE
IN DEGREES FOR A SERIES OF DECADES AT DIFFERENT POINTS ON
THE NORTH AMERICAN CONTINENT. THE PLUS (+) SIGN PRE-
FIXED TO A NUMBER INDICATES A WESTERN DECLINATION, AND
THE MINUS (−) SIGN AN EASTERN DECLINATION.

Year (Jan. 1)	Halifax, N.S.	Cambridge, Mass.	New York City.	Washington, D.C.	Erie, Pa.	New Orleans, La.	San Francisco, Cal.	Sitka, Alaska.
1700		+9.80	8.50					
1710		9.20	8.00					
1720		8.70	7.60			−3.40		
1730		8.30	7.20			−3.70		
1740		7.90	6.60			−4.10		
1750	+12.5	7.50	5.90			−4.70		
1760	13.0	7.20	5.20			−5.30		
1770	13.7	7.00	4.60			−5.90		
1780	14.4	6.90	4.40			−6.50		
1790	15.1	6.90	4.29	−0.10	−0.70	−7.00	−12.80	
1800	15.9	7.10	4.28	0.00	−0.70	−7.50	−13.40	−26.12
1810	16.7	7.50	4.30	+0.30	−0.60	−7.90	−13.90	−27.11
1820	17.4	8.00	4.47	0.60	−0.30	−8.10	−14.42	−27.89
1830	18.1	8.64	4.91	1.00	+0.03	−8.20	−14.92	−28.48
1840	18.7	9.33	5.59	1.49	0.44	−8.14	−15.38	−28.88
1850	19.3	10.03	6.34	1.99	0.91	−7.94	−15.78	−29.08
1860	19.8	10.67	6.96	2.47	1.39	−7.61	−16.11	−29.08
1870	20.1	11.51	7.43	2.90	1.87	−7.15	−16.36	−28.88
1880	+20.3	+11.63	+7.84	+3.26	+2.31	−6.62	−16.52	−28.50

NOTE. "The west declination of the magnetic needle at Cambridge,
Mass., for the beginning of 1884 is + 11.760°, and increased annually, at
the epoch 1880, 0.0347°. The minimum west declination at this place
occurred about the epoch 1782, 0.7°. The magnetic dip at this place is
74°."—E. C. PICKERING, Harvard College Observatory.

MEAN INDICES OF REFRACTION AND DISPERSIONS OF SEVERAL
SUBSTANCES.

www.ingramcontent.com/pod-product-compliance
Lightning Source LLC
Chambersburg PA
CBHW021952190326
41519CB00009B/1230